Underwater Acoustics and Ocean Dynamics

Lisheng Zhou · Wen Xu
Qianliu Cheng · Hangfang Zhao
Editors

Underwater Acoustics and Ocean Dynamics

Proceedings of the 4th Pacific Rim Underwater Acoustics Conference

ZHEJIANG UNIVERSITY PRESS
浙江大学出版社

Springer

Editors
Lisheng Zhou
Hangzhou Applied Acoustics Institute
Hangzhou, Zhejiang
China

Qianliu Cheng
Hangzhou Applied Acoustics Institute
Hangzhou, Zhejiang
China

Wen Xu
College of Information Science & Electronic Engineering
Zhejiang University
Hangzhou, Zhejiang
China

Hangfang Zhao
College of Information Science & Electronic Engineering
Zhejiang University
Hangzhou, Zhejiang
China

ISBN 978-981-10-2421-4 ISBN 978-981-10-2422-1 (eBook)
DOI 10.1007/978-981-10-2422-1

Jointly published with Zhejiang University Press

Library of Congress Control Number: 2016948114

Printed on acid-free paper

This Springer imprint is published by Springer Nature
The registered company is Springer Science+Business Media Singapore Pte Ltd.

Preface

The 4th Pacific Rim Underwater Acoustics Conference (PRUAC) was held in Hangzhou, China during 9–11 October 2013. Thanks to all the participants, it turned out to be one of the most exciting conferences around the Pacific focused on the ocean, with the theme "Underwater Acoustics and Ocean Dynamics."

This conference was jointly hosted by Hangzhou Applied Acoustics Research Institute and Zhejiang University. The sponsors also included Acoustical Society of China, Science and Technology on Sonar Laboratory, Acoustical Society of America, Canadian Acoustical Association and Acoustical Society of Korea. The objective of this conference was to provide a forum for active researchers to discuss state-of-the-art developments in underwater acoustics. It brought together scholars, scientists, and engineers from numerous countries to exchange ideas and stimulate future research.

The proceedings are a collection of most of the scientific papers and reviews that were presented at the 4th PRUAC conference. The volume is comprised of 16 presented lectures covering a variety of topics in three sessions, including acoustical oceanography, underwater acoustic communication, and vector sensors and target detection. These lectures were made by distinguished researchers from a variety of countries, including the United States, China, Canada, Korea, and Russia.

We extend our sincere gratitude to all the attendees of the conference, with special thanks to the invited speakers, conference committee members, and those who have provided support for the conference and proceedings.

With the PRUAC 2015 conference coming soon in Russia, we look forward to meeting all our fellow researchers there.

Hangzhou, China

Lisheng Zhou
Wen Xu
Qianliu Cheng
Hangfang Zhao

Organization Committee

Scientific Program Committee

Xianyi Gong, Hangzhou Applied Acoustics Research Institute, China
Renhe Zhang, Institute of Acoustics, China
Robert Spindel, University of Washington, USA
Ross Chapman, University of Victoria, Canada
Peter Dahl, University of Washington, USA
Jungyul Na, Hanyang University, Korea
Wen Xu, Zhejiang University, China
Xiaowen Cui, Hangzhou Applied Acoustics Research Institute, China
Fenghua Li, Institute of Acoustics, China

Organizing Committee

Lisheng Zhou, Co-chair, Hangzhou Applied Acoustics Research Institute, China
Wen Xu, Co-chair, Zhejiang University, China
Shuanping Du, Hangzhou Applied Acoustics Research Institute, China
Hangfang Zhao, Zhejiang University, China

Sponsors

Hangzhou Applied Acoustics Research Institute, China
Zhejiang University, China
Acoustical Society of China

Science and Technology on Sonar Laboratory, China
Acoustical Society of America
Canadian Acoustical Association
Acoustical Society of Korea

Contents

Inference of Sound Attenuation in Marine Sediments from Modal Dispersion in Shallow Water 1
N. Ross Chapman and Juan Zeng

Counterintuitive Results in Underwater Acoustic Communications 11
Daniel Rouseff

Comparisons of Methods for Numerical Internal Wave Simulation in Long-Range Acoustical Propagation 19
Rex K. Andrew

Acoustic Data Assimilation: Concepts and Examples 27
Liling Jin, Wen Xu, Jiamin Huang and Jianlong Li

The Preliminary Results of a Single-Hydrophone Geoacoustic Inversion for Data Collected at the Sea of Japan 37
Pavel S. Petrov

Measurements of Ultrasound Attenuation of Suspended Particles with Various Size Distributions 45
Changil Lee, Jee Woong Choi and Jungyul Na

A Low Complexity Multichannel Adaptive Turbo Equalizer for a Large Delay Spread Sparse Underwater Acoustic Channel 49
Yanbo Wu and Min Zhu

A Turbo Equalization Based on a Sparse Doubly Spread Acoustic Channels Estimation 57
Zibin Yu, Hangfang Zhao, Wen Xu and Xianyi Gong

Design and Testing of Underwater Acoustic Communications for an AUV 63
Hongtao Zhang, Zhijie Yue, Zhe Xie and Zhongkang Wang

Research of Axis Mismatches Between Pairs of Sensitive Elements of Underwater Acoustic Velocity Gradient Sensors 71
Xinyi Sun, Desen Yang, Lianjin Hong, Shengguo Shi and Hongkun Zhou

A Passive Fathometer Technique for Bottom Profiling Using Ambient Noise 81
Junghun Kim, Jee Woong Choi and Jungyul Na

Target Motion Parameter Estimation for LOFARgrams Based on Waveguide Invariants 85
Jian Li, Guangjie Han, Defu Zhou, Kai Tang and Qingbang Han

Source Localization by Maximizing the Longitudinal Correlation Using Waveguide Invariant Theory 93
Xian Zhu and Huiliang Ge

Selective Detection and Localization by Decomposition of a Subrank Time Reversal Operator 99
Chunxiao Li, Mingfei Guo and Huancai Lu

Sounds of Undersea Gas Leaks 107
Alexey O. Maksimov, Boris A. Burov, Alexander S. Salomatin and Denis V. Chernykh

Multi-AUV Localization for an Underwater Acoustic Sensor Network 117
Bo Li, Yuanxin Xu and Wen Xu

Inference of Sound Attenuation in Marine Sediments from Modal Dispersion in Shallow Water

N. Ross Chapman and Juan Zeng

Abstract Attenuation of sound in the seabed plays an important role in predicting transmission loss in shallow water waveguides. Methods to invert the attenuation from low-frequency acoustic field data include time-frequency techniques that make use of modal dispersion. Since modal separation improves as a sound signal that propagates to longer ranges, most of the inversion methods based on modal dispersion were carried out with long range data. Recently a time-warping signal processing technique was introduced that enables high resolution of modes at relatively short ranges. Time-warping involves an axis transformation that transforms the original time-frequency relationship of the modes to a new domain in which the modes are approximately tonal and are well resolved. This paper shows that the inversion can be carried out directly in the time-warped domain, and extends the work to estimate low-frequency seabed attenuation.

Keywords Geoacoustic inversion · Time-warping transform · Seabed attenuation · Modal dispersion

1 Introduction

In shallow water, it is well known that the geoacoustic properties of the seabed have a significant impact on sound propagation at low frequencies (<1 kHz). Inversion methods for estimation of seabed model parameters from acoustic field data have been developed by many researchers, and benchmark exercises to compare the inversion performance have shown that they provide realistic results for the sound

N.R. Chapman (✉)
School of Earth and Ocean Sciences, University of Victoria, Victoria
BC V8P5C2, Canada
e-mail: chapman@uvic.ca

J. Zeng
Laboratory of Underwater Environment, Institute of Acoustics,
CAS, Beijing 100190, China

L. Zhou et al. (eds.), *Underwater Acoustics and Ocean Dynamics*
DOI 10.1007/978-981-10-2422-1_1

speed profile in marine sediments [1]. However, estimation of low-frequency sound attenuation in marine sediments remains a significant experimental challenge.

The most comprehensive summaries of attenuation measurements at low frequencies are due to Holmes and Carey [2] and Zhou et al. [3] who compiled results from many different types of experiments. Inversion methods for attenuation can be divided into three categories according to the type of experimental data that were used in the inversions: (1) modal techniques that use dispersion or wave number analysis to separate propagating modes [4–8], (2) transmission loss (TL) techniques that use TL versus range measurements at multiple frequencies with narrowband or broadband impulsive sources [8], and (3) reflection coefficient techniques that measure the angle-dependent reflection coefficient at different frequencies [9].

Because of the strong dispersion of propagating modes in shallow water, the techniques based on modal dispersion offer significant promise for the estimation of seabed attenuation. However, most of the inversion methods based on dispersion phenomena made use of long range data ($> \sim 100$ water depths), because modal separation improves with range and the dispersion curve and modal amplitude ratios can be extracted directly by time-frequency analysis of the received signal [6, 7]. The disadvantage of using long-range data is the impact of the range-dependent environment, including variations of water depth and changes in the type of sediment material.

For close range data, special signal processing techniques known as warping have recently been introduced that enable high resolution of the propagating modes, and there have been many successful applications in the analysis of the underwater signals. Iaona et al. [10] used the warping operator to analyze the signals emitted by marine mammals. Bonnel and Gervais [11] used it to extract the arrival times of different modes at different frequencies and the mode functions from the received pressure signal emitted by an air-gun source. Gao et al. [12] used an invariant-based warping operator to remove the dispersion effect from close-range received data.

The warping operation transforms the original signal into a new time and frequency space in which the modes are well-resolved tones. Previous work by Bonnel and Chapman [13] has shown that time-warping of broadband signals in shallow water provides estimates of modal dispersion curves that can be used effectively for inverting sound speed and density in marine sediments. However, the inversion required transforming the resolved modes back into the original time-frequency domain.

This paper shows that the inversion can be carried out in the time-warped domain, and extends the initial work to estimate low-frequency seabed attenuation. The group velocities and modal amplitudes that are used in a two-stage inversion are extracted directly from the spectrum of the warped signal. Sound speed and density of the seabed are inverted from the dispersion curve in the first stage, and then the estimated values are used as prior information to invert attenuation from the normalized modal amplitudes. The method is applied to invert attenuation using short-range data from an experiment with impulsive broadband sources carried out in the Yellow Sea off the east coast of China.

The paper is organized as follows. The theory of inversion of modal dispersion in the warped domain is briefly reviewed in the next section, and a simulation is carried out to demonstrate the feasibility of the approach. The experiment is described, and the results of the two-stage inversion are presented. The estimated values of attenuation are compared with results from other experiments in the Yellow Sea. The last section summarizes the paper.

2 Inversion by Time-Warping

2.1 *Mode Relationships*

The theoretical development of the time-warping transform has been reported previously [13, 14] and only an outline of the main points will be given here. In a shallow water waveguide, low-frequency sound propagation is dispersive. The dispersion relationship of mode m satisfies

$$t_m(\omega) = \frac{r}{v_g^m(\omega)} \tag{1}$$

where t_m is the arrival time of the component at frequency ω of mode m at range r, and v_g^m is its group velocity. The group velocity is related to the geoacoustic properties of the ocean waveguide by [15]

$$\frac{1}{v_g^m(\omega)} = \frac{\omega}{k_m(\omega)} \int_D^\infty \frac{\rho_b(z)}{c_b^2(z)} |\Psi_m(z)| \mathrm{d}z + \frac{\omega}{k_m(\omega)} \int_0^D \frac{\rho(z)}{c^2} |\Psi_m(z)| \mathrm{d}z \tag{2}$$

where k_m and Ψ_m are the horizontal wave number and mode function of mode m, respectively, D is the water depth, ρ_b, ρ and c_b, c are the densities and sound speeds in the ocean bottom and the water, respectively. The normalized amplitude of the mth mode can be expressed as

$$A_m(\omega) = 1 \Big/ \sqrt{\sum_{n=1}^{M} \left| \frac{\Psi_n(z_s)\Psi_n(z_r)}{\Psi_m(z_s)\Psi_m(z_r)} \right|^2 \left| \frac{k_m}{k_n} \right|^2 \mathrm{e}^{-2(\beta_n - \beta_m)r}} \tag{3}$$

where M is the total number of modes, β_m is the mode attenuation that is given by [15]

$$\beta_m = \frac{\omega}{k_m(\omega)} \int_D^\infty \frac{\alpha_b(z)}{c_b(z)} \rho_b(z) |\Psi_m(z)|^2 \mathrm{d}z + \frac{\omega}{k_m(\omega)} \int_0^D \frac{\alpha}{c} \rho |\Psi_m(z)|^2 \mathrm{d}z \tag{4}$$

where α_b and α are the attenuations in the ocean bottom and water, respectively.

From Eq. (2), it is clear that the group velocity is sensitive only to the density and sound speed in the bottom, and from Eq. (4), the mode attenuation and thus the normalized mode amplitude is sensitive to all three bottom parameters, namely attenuation, sound speed, and density. Our inversion is therefore staged in two parts: the sound speed and density of the sediment bottom are obtained first from inversion of the modal group velocity, and then the sound attenuation in the bottom is obtained by inversion of the normalized mode amplitude, using the results of the first inversion as prior knowledge of sound speed and density.

2.2 *Time-Warping*

Application of the time-warping transform follows the development in Refs. [13, 14]. We assume an ideal waveguide, for which the warping function has the form

$$h(t) = \sqrt{t^2 + t_r^2} \tag{5}$$

where $t_r = r/c$, the travel time from the source to the receiver, which is conveniently the travel time of the highest frequency component of the first mode. Although Eq. (5) is defined for an ideal waveguide, it is a robust approximation, and can be applied to most low-frequency shallow water environments.

The relationship between the warped frequency ω_w and the original time is [10]

$$\omega_w = \omega_0\sqrt{1 - (t/t_r)^2} \tag{6}$$

where ω_0 is the center frequency of the original signal. According to Eq. (6), the frequency of the signal has been changed after the warping transform, and is a function of time. The relationship between the arrival time of the mth mode $t_m(\omega_0)$, and the warped frequency ω_w^m corresponding to the mode is thus,

$$t_m(\omega_0) = t_r/\sqrt{1 - (\omega_w^m/\omega_0)^2} \tag{7}$$

Substituting Eq. (7) into Eq. (1), the group velocity can be obtained directly,

$$v_g^m = c\sqrt{1 - (\omega_w^m/\omega_0)^2} \tag{8}$$

Using Eq. (8) the group velocity curve is extracted directly from the spectrum of the warped signal without the need to transform back into the original time-frequency domain. The relationship is exact for an ideal waveguide, and is approximately true for a real waveguide. Since the warping transform conserves energy, the modal amplitude ratios can also be extracted from the spectrum of the

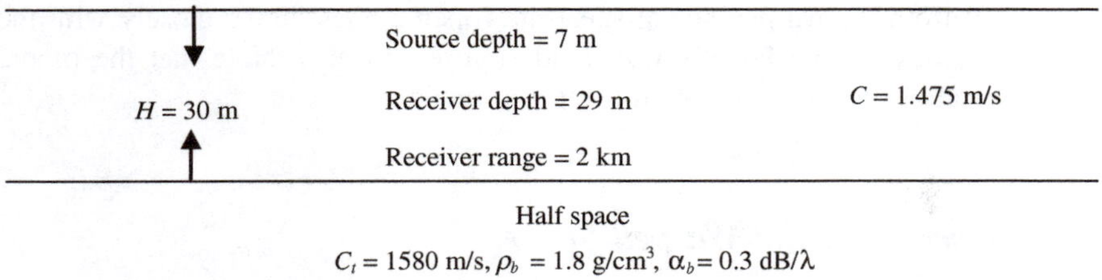

Fig. 1 Model parameters of the Pekeris waveguide. H water depth; C sound speed; C_b sound speed in the bottom half space; ρ_b the density; α_b the attenuation

warped signal. In the inversion, the modal amplitudes can be normalized over all the resolved modes to generate a more stable cost function that is robust to errors in the amplitude of any single mode [14].

A simulation using the shock pulse of the shot as the sound source was carried out to demonstrate the feasibility of the approach. The center frequency of the simulated signal was 500 Hz and the band was 397–630 Hz. The model parameters of the Pekeris waveguide environment are shown in Fig. 1. The comparison of the calculated group velocities and modal amplitude ratios (calculated using KRAKENC [16]) with those determined from the spectrum of the warped signal is shown in Fig. 2. The left panel shows the warped signal spectrum for the centre frequency, the middle panel shows the mode amplitude ratios (with respect to mode 2) and the right panel shows the group velocities. Figure 2 indicates that the estimated values

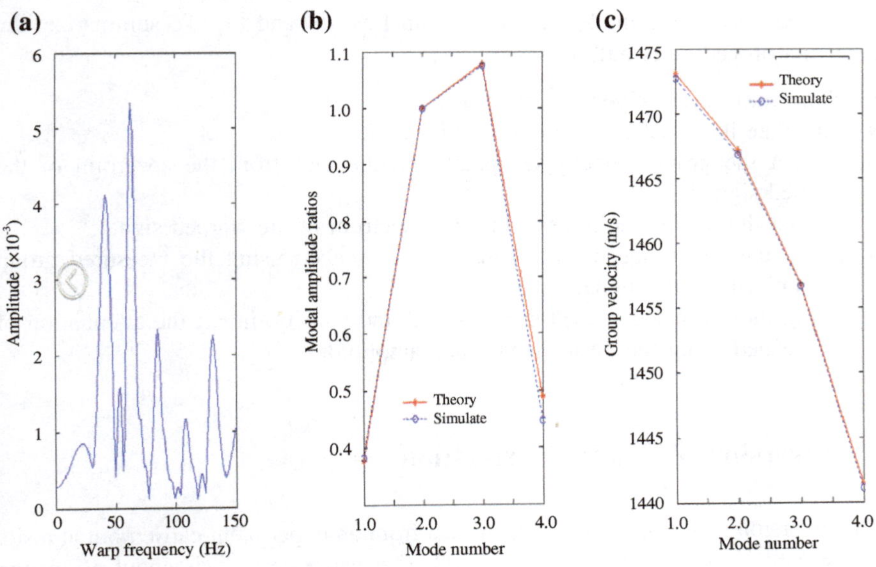

Fig. 2 Spectrum of the warped signal (**a**), modal amplitude ratios (**b**) and group velocities (**c**)

determined from the warped signal spectrum (open circles) agree closely with the expected values for the Pekeris waveguide (closed circles). Note that the modal amplitude ratios are calculated with respect to mode 2.

2.3 *Inversion in the Warped Space*

With the estimated group velocities and modal amplitude ratios, the inversion can be carried out using standard methods. The cost function for the inversion of the sound speed and density is

$$C_v(\theta) = \sum_{m,n=1}^{M,N} \left[\widehat{v}_g^m(\omega_n) - v_g^m(\omega_n, \theta) \right]^2 \tag{9}$$

and the cost function for the inversion of sound attenuation in the seabed is

$$C_A(\alpha) = \sum_{m=1}^{M} \left[\widehat{A}_m(\omega_n) - A_m(\omega_n, \alpha) \right]^2 \tag{10}$$

In Eqs. (9) and (10), M is the number of modes, N is the number of the frequencies in the broadband sound signal, the vector θ is the set of inversion parameters, $\widehat{v}_g^m(\omega_n)$ is the estimated group speed from the experimental data, and $v_g^m(\omega_n, \theta)$ is the calculated group speed according to Eq. (2). $\widehat{A}_m(\omega_n)$ is the normalized mode amplitude estimated from the experimental data, and $A_m(\omega_n, \alpha)$ is the calculated normalized mode amplitude from Eqs. (3) and (4). To summarize, the inversion procedure is outlined as follows:

- Time-warp the received signal;
- Calculate the spectrum of the warped signal;
- Extract the group velocity at specific frequencies from the spectrum of the warped signal;
- Extract the mode amplitudes from the spectrum of the warped signal;
- Invert the sound speed and density of the seabed using the measured group velocities over the signal band;
- Using the estimated seabed sound speed and density, invert the attenuation of the seabed using the normalized mode amplitudes.

3 Inversion of Sound Attenuation

The inversion is applied to broadband data from an experiment carried out at a site in the Yellow Sea in the winter of 2002. The water depth was about 30 m, the sound speed profile of the water was almost constant at ~1475 m/s, and the ocean

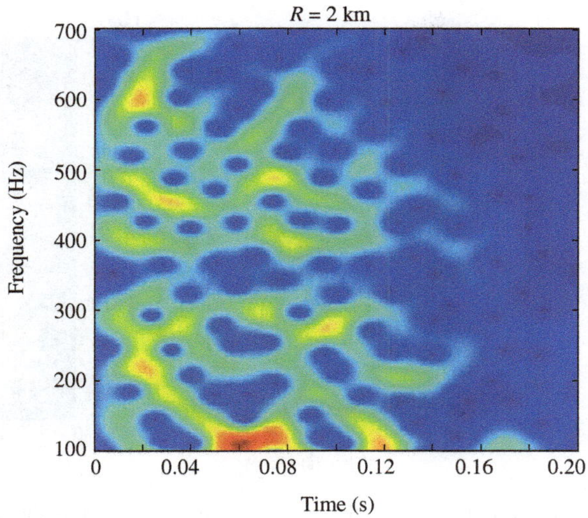

Fig. 3 Spectrogram of the 38*g* shot signal received at the deepest array hydrophone at a range of 2 km. The bubble pulse of the charge contaminates the signal after ~60 ms

bottom between the source and the receiver was essentially range independent. The sound sources were small explosive charges (38 g) that were detonated at ~7 m, and the shot signals were received at a vertical line array of 30 sensors with an interval between the sensors of 1 m. The sediment bottom at the site was a mixture of sand-silt-clay.

The data received by the deepest sensor (about 29 m) at a range of 2.1 km were used in the inversion. The spectrogram of the shot signal over the band 100–700 Hz shown in Fig. 3 indicates that there is significant intra- and inter-mode interference, and clearly the modes are not well resolved. In addition, the bubble pulse of the charge arrives about 62 ms after the onset of the signal, and contaminates the data at longer time. For this reason, the warping transform was applied to a portion of the signal that was time-windowed at 60 ms.

The spectrograms of the time-windowed signal before and after time-warping are shown in Fig. 4. There are eight modes resolved by the warping transform and these were used in the first stage of the inversion to estimate sound speed and density in the bottom sediment. A Pekeris waveguide model was assumed based on the results of sub-bottom sonar surveys that indicated no significant layering in the sediment in the vicinity of the experimental site. Since the number of unknowns was small, the inversion was cast as a grid search for three unknown model parameters: the water depth, the sound speed, and density of the sediment bottom, and all other parameters were assumed to be the values measured in the experiment. The search ranges and estimated values of the unknown parameters are listed in Table 1. For the estimated sound speed, the ratio of sediment/water sound speed is 1.07. The curves shown in Fig. 4a represent the calculated modal arrival time for the first four modes using the estimated model parameters.

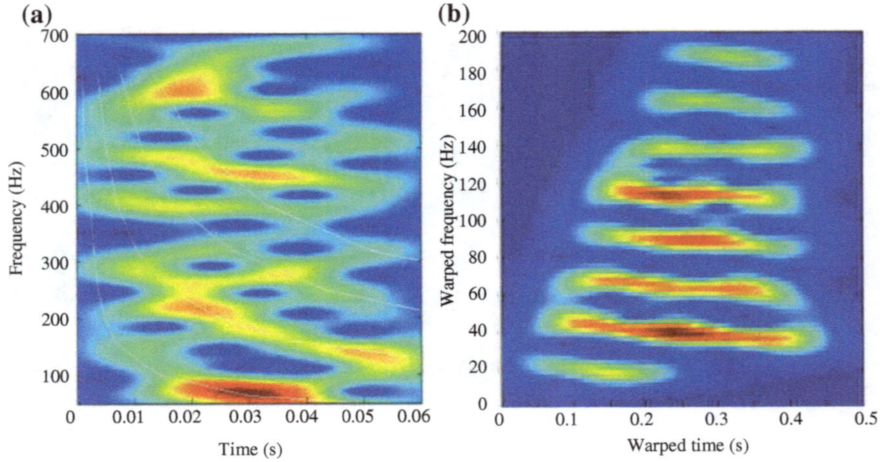

Fig. 4 Spectrograms of the time-windowed shot signal in the real time-frequency (**a**) and warped (**b**) domains. The curves through the first four modes in the original time-frequency spectrogram are calculated based on the estimated model parameters

Table 1 The search range and inverted results of the geoacoustic model parameters

Parameter	Unit	Range	Step	Inverted result
Water depth	m	[28, 32]	0.2	29.7
Sound speed	m/s	[1500, 1800]	5.0	1580.0
Density	g/cm^3	[1.0, 2.2]	0.1	1.7

Sound attenuation was inverted from Eq. (10) over the band 125–500 Hz using the estimated values of sound speed and density from Table 1. The inversion was a grid search over the range [0, 1.6] dB/λ with step size of 0.1 dB/λ. The results listed in Table 2 are consistent with a nonlinear frequency dependence of attenuation given by $\alpha(f) = \alpha_0(f/f_0)^\beta$ with $\alpha_0 = 0.55$ and $\beta = 1.9$. These values are slightly higher than those reported previously [3, 17]. The significant aspects of our results are: they were obtained using more information from higher order modes at short range (~60 water depths), and thus represent an average over about the first 10 m of sediment below the sea floor. The results from other experiments were obtained from lower order modes (modes 1 and 2), and from very long range data (>500 water depths) [3, 6, 7].

Table 2 Estimates of the seabed attenuation

Frequency (Hz)	125	160	200	250	320	400	500
Attenuation (dB/m)	0.0134	0.0334	0.0352	0.0522	0.0567	0.109	0.174

4 Conclusion

This paper presents an application of time-warping for inversion of sound attenuation in marine sediments. The estimated geoacoustic model parameters are consistent with those expected for the sand-silt-clay sediment material, and the attenuation displays nonlinear frequency dependence over the low-frequency band below 500 Hz.

References

1. Chapman, N.R., Chin-Bing, S.A., King, D., Evans, R.: Benchmarking geoacoustic inversion methods for range dependent waveguides. IEEE J. Oceanic Eng. **28**(3), 320–330 (2003)
2. Holmes, J.D., Carey, W.M.: Nonlinear frequency-dependent attenuation in sandy sediments. J. Acoust. Soc. Am. **121**(5), EL218–EL222 (2007)
3. Zhou, J.X., Zhang, X.Z., Knobles, D.P.: Low-frequency geoacoustic model for the effective properties of sandy sea bottoms. J. Acoust. Soc. Am. **125**(5), 2847–2866 (2009)
4. Tindle, C.T.: Attenuation parameters from normal mode measurements. J. Acoust. Soc. Am. **71**(5), 1145–1148 (1982)
5. Rogers, P.H., Zhou, J.X., Zhang, X.Z., Li, F.: Seabottom acoustic parameters from inversion of Yellow Sea experimental data. In: Caiti, A., Hermand, J.P., Jesus, S.M., Porter, M.B. (eds.) Experimental Acoustic Inversion Methods for Exploration of the Shallow Water Environment, pp. 219–234. Kluwer (2000)
6. Zhou, X., Zhang, X.Z., Rogers, P.H., Jarzynski, J.: Geoacoustic parameters in a stratified sea bottom from shallow-water acoustic propagation. J. Acoust. Soc. Am. **82**(6), 2068–2074 (1987)
7. Wan, L., Zhou, J.X., Rogers, P.H.: Low-frequency sound speed and attenuation in sandy sea bottom from long-range broad-band acoustic measurements. J. Acoust. Soc. Am. **128**(2), 578–589 (2010)
8. Carey, W.M., Evans, R.E.: Frequency dependence of sediment attenuation in two low-frequency shallow-wateracoustic experimental data sets. IEEE J. Oceanic Eng. **23**(4), 439–447 (1998)
9. Cole, B.: Marine sediment attenuation and ocean-bottom-reflected sound. J. Acoust. Soc. Am. **38**(2), 291–297 (1965)
10. Ioana, C., Quinquis, A., Stephan, Y.: Feature extraction from underwater signal using time-frequency warping operators. IEEE J. Oceanic Eng. **31**(3), 628–645 (2006)
11. Bonnel, J., Gervaise, C.: Modal depth function estimation using time-frequency analysis. J. Acoust. Soc. Am. **130**(1), 61–71 (2011)
12. Gao, D.Z., Wang, N., Wang, H.Z.: A dedispersion transform for sound propagation on shallow water waveguide. J. Comp. Acoustics **18**(3), 245–258 (2010)
13. Bonnel, J., Chapman, N.R.: Geoacoustic inversion in a dispersive waveguide using warping operators. J. Acoust. Soc. Am. **130**(2), EL101–EL107 (2011)
14. Zeng, J., Chapman, N.R., Bonnel, J.: Inversion of seabed attenuation using time-warping of close range data. J. Acoust. Soc. Am. **134**(5), EL394–EL399 (2013)
15. Koch, R.A., Penland, C., Vidmar, P.J., Hawker, K.E.: On the calculation of normal mode group velocity and attenuation. J. Acoust. Soc. Am. **73**(3), 820–825 (1983)
16. Porter, M.B. The KRAKEN Normal Mode Program. Report No. NRL/MR/5120-92-6920. Naval Research Laboratory, Washington, DC (1992)
17. Li, Z.L., Zhang, R.H.: A broadband geoacoustic inversion scheme. Chin. Phys. Lett. **24**, 1100–1103 (2004)

Counterintuitive Results in Underwater Acoustic Communications

Daniel Rouseff

Abstract Underwater wireless communication using acoustic signals is a difficult problem and progress in finding robust solutions has been disappointing. Methods employed successfully in terrestrial wireless communications have not always transitioned successfully to underwater scenarios. An engineer's intuition developed in solving the terrestrial problem may actually become a hindrance to solving the underwater problem. In the present work, several seemingly counterintuitive experimental results are examined: communications performance can be better when the range is longer rather than shorter, when the sea surface is rough rather than calm, when the bathymetry is undulating rather than flat. Physics-based explanations for the observed results are developed. A physicist's intuition, however, also may fail when trying to develop useful models. A seemingly counterintuitive fact is that acoustic paths that undergo incoherent reflection from a rough sea surface can be shown experimentally to be useful for coherent communications. The requirements for a proper physics-based model are sketched.

Keywords Underwater acoustic communications · Rayleigh parameter · Equalization

1 Introduction

In 2000, Kilfoyle and Baggeroer [1] published a comprehensive review of progress in underwater acoustic communications since 1982. They noted how early work had concentrated on incoherent methods, but since the early 1990s there had been

---

D. Rouseff (✉)
Applied Physics Laboratory, University of Washington, Seattle, WA 98105, USA
e-mail: rouseff@apl.washington.edu

D. Rouseff
Northwest Electromagnetics and Acoustics Research Laboratory,
Department of Electrical Engineering, Portland State University,
Portland, OR 9720, USA

L. Zhou et al. (eds.), *Underwater Acoustics and Ocean Dynamics*
DOI 10.1007/978-981-10-2422-1_2

numerous publications on coherent systems. Coherent systems make better use of the available bandwidth and so are capable of much higher data rates. With coherent modulation and complex receiver structures, systems deployed at sea had achieved a data rate time's range product of nearly 40 km × kbit.

While research in underwater acoustic communications has certainly advanced since 2000, it is not unfair to say that the results have been somewhat disappointing. This is not to say that there has not been progress in equalizer design, or that there have not been several high quality scientific experiments. What has been lacking, though, is robustness; a prototype system might work well on one day or at one location, but then fail on another day or at another location. This lack of robustness has hampered transition from prototype systems used in demonstration experiments to viable commercial products.

The reasons why underwater acoustic communication is so difficult are well known: the communications channel is lossy, inhomogeneous, rapidly varying, and vulnerable to Doppler shifts [1]. While these difficulties are well known in the abstract, when they actually manifest themselves in experimental data the results are still often surprising, seemingly counterintuitive: communication performance can get better and be more stable as the source–receiver distance is increased, small changes in source or receiver depth can have a major effect on performance while changing the transmission power can have no effect at all, acoustic paths that are incoherently scattered by a rough sea surface can still be useful for coherent communication.

The premise of this paper is that many of these seemingly counterintuitive results have well-understood physical bases and can be adequately modeled. By understanding these bases, one can develop a better intuition, develop better performance prediction models, and perhaps, ultimately design better communications systems.

This paper reviews seemly counterintuitive results from three experiments in which the author participated: the 2000 Puget Sound Passive Phase Conjugation Experiment [2–5], the 2003 Kauai Experiment (KauaiEx03) [6–8], and the 2009 Cooperative Array Performance Experiment (CAPEx09) [9, 10]. In each case, relatively simple, physics-based models can reproduce the experimental results. Even seemingly complicated effects like scattering from the spatially rough and temporally varying sea surface can yield to fairly simple modeling efforts.

2 The Rayleigh Parameter

Oceanographic processes operate over a wide range of spatial and temporal scales. The oceanographic process with particular relevance for acoustic communication is waves on the sea surface. This is true for two interrelated reasons. First, the height of the waves can be large compared to the acoustic wavelength. As a result, the phase of a surface-reflected path will depend strongly on where the sea surface wave is in its cycle. This is important for coherent communications techniques where a signal's information is encoded in its phase. Second, the period of the surface wave is often comparable to the duration of a communications packet. The acoustic channel cannot therefore be regarded as stationary over time scales on the order of the wave period. To develop an intuition for representative numbers, a fully developed sea driven by a 10 m/s wind has waves with a dominant period of about 7 s. An equalizer would have to adapt to an environment changing at these time scales.

A useful quantity for describing acoustic paths that interact with the sea surface is the Rayleigh parameter. Assume that there are multiple acoustic paths between a source sending a communications signal with center frequency f and a receiver. The Rayleigh parameter for a path incident on the rough sea surface at grazing angle θ is

$$P = 2k\sigma \sin(\theta) \tag{1}$$

where $k = 2\pi f/c$ is the acoustic wavenumber with c the sound speed in the water at the air–sea interface, and σ the standard deviation of the height of the waves.

As observed by Eckert, a Rayleigh parameter $P = 2$ is sufficient to cause a loss in the coherent reflection of 15 dB [11]. This energy is not actually lost, but rather redistributed into the incoherent field scattered off the sea surface. Figure 1 is an example showing the effects of incoherent reflection off the sea surface. Shown is the time-evolving channel impulse response (CIR) for the KauaiEx03 experiment at a range of 1 km. Specifically, linear frequency-modulated chirp signals, 50 ms in

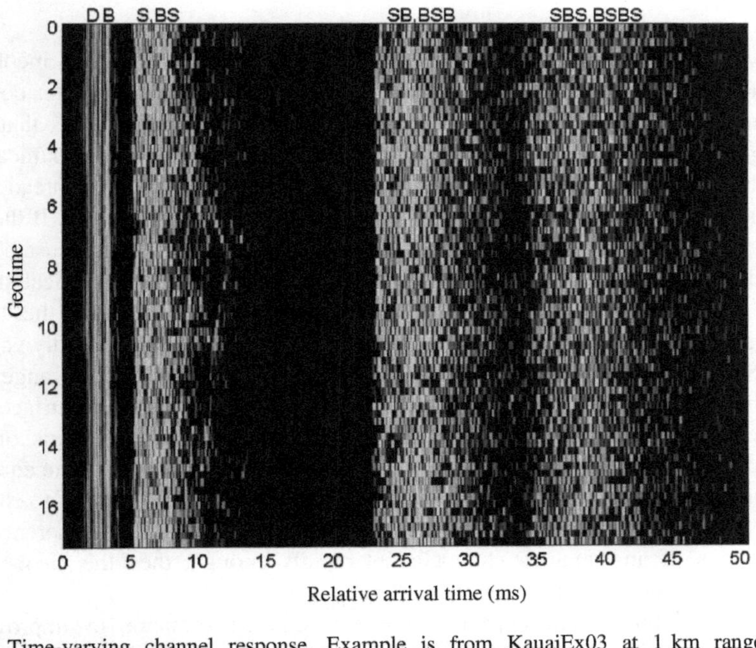

Fig. 1 Time-varying channel response. Example is from KauaiEx03 at 1 km range from transmitter [2]. Direct (*D*), bottom-bounce (*B*), surface-bounce (*S*), and multiple-bounce paths are labeled. The time spread for each path reflected off the sea surface is consistent with that having a Rayleigh parameter greater than one. The dynamic range is 25 dB

duration and spanning the 8–16 kHz band of the communications sequences, were transmitted every 250 ms for 20 s. The output of a matched filter for consecutive chirps is then stacked and displayed as shown. Because the acoustic source was located near the seabed, the acoustic arrivals come in pairs. The first pair, with the direct (D) and bottom-bounce (B) paths, does not interact with the sea surface. These paths show little time spread and are stable over the 20 s. The second pair has the surface-bounce path (S) and the path that reflects first off the bottom and then the sea surface (BS). This second pair of arrivals is less distinct than the first. Using environmental data collected concurrently, the Rayleigh parameter for this pair was shown to be 4.4 [8]. The interpretation is clear: acoustic paths that get reflected off the sea surface and have a large Rayleigh parameter get smeared out in time. It follows that the larger the Rayleigh parameter, the more pronounced this smearing becomes. This can be observed in Fig. 1 for later arriving paths labeled as the (SB, BSB) pair and the (SBS, BSBS) pair. Note that beginning with the latter pair, the paths have multiple reflections from the sea surface. From the standpoint of Eq. (1), later arriving paths reflect off the sea surface at a steeper angle and so have larger Rayleigh parameters. For the (SB, BSB) pair, $P = 8.3$ and for the (SBS, BSBS) pair, $P = 10.6$.

3 Communications Performance

Intuitively, one might expect that acoustic paths that have undergone incoherent reflection and have $P > 1$ could not then be used for coherent acoustic communication. The experimental observation is to the contrary: some paths that have undergone incoherent reflection are, in fact, useful for coherent communications. What limits communications performance is not necessarily the time spread in the CIR like what is evident by looking at a horizontal slice through Fig. 1. If the time spread was unchanging, an equalizer could be easily designed to compensate for it. Instead, it is the time variability in the CIR as is observed along a vertical slice in Fig. 1 that limits performance. The later arriving paths that have the largest Rayleigh parameters and are the most time spread are also the most rapidly varying. For the particular equalizer used in KauaiEx03, the best performance at range 1 km was observed when all paths with at most one reflection off the sea surface were retained and treated as useful signal rather than interference [8]. This optimal performance could only be achieved, though, when the parameters of the equalizer were updated sufficiently often. If an equalizer can adapt rapidly enough to changes in the environment, incoherently reflected paths can be used for coherent communications. If an equalizer cannot adapt rapidly enough, then the incoherently reflected paths act as inter-symbol interference.

Intuitively, one might expect communications performance to improve by transmitting louder and thereby increasing the signal strength. This is not necessarily correct. In this regard, it is important to distinguish between additive noise and inter-symbol interference. In principle, one could indeed combat additive noise simply by transmitting louder. Transmitting louder, however, does not defeat interference from incoherently reflected paths that are uncompensated. By transmitting louder, both the desired signal and the undesired interference are magnified with no resulting improvement in communications performance. For the linear equalizer used at CAPEx09, it was shown how doubling the transmit voltage had no effect on communications performance at range 1 km [10]. For the nonlinear equalizer used at KauaiEx03, individual surface-reflected paths were either signal or interference depending on the equalizer's adaptation rate [8].

It is sometimes possible to position a communications system in such a way as to have the ocean environment help rather than hinder operations. This was illustrated by data collected during the 2000 Puget Sound Experiment [2]. The source and receiver were separated by 1.6 km, both in water 30 m deep. The bathymetry in between was rugged with a crevasse 80 m deep. Communications performance was error free over 5 s without updating the equalizer. A ray trace suggested that the crevasse was serving to strip out the acoustic rays traveling with steep grazing angles and so large Rayleigh parameters. The result was an extremely stable communications channel.

Just as the bathymetry was important in Puget Sound, the sound speed profile had a significant effect on communications performance in CAPEx09. Warm surface-layer water overlaid much cooler water below causing a sharp sound speed gradient. Figure 2a shows the profile where the sound speed contrast between the top and bottom is almost 50 m/s. The result shown in Fig. 2b is strong refraction of the acoustic rays emanating from the source at a depth of 30 m. Figure 2b also highlights four ranges where the receiving arrays were positioned when communications sequences were transmitted: 100, 250, 500, and 1 km. The sharp sound

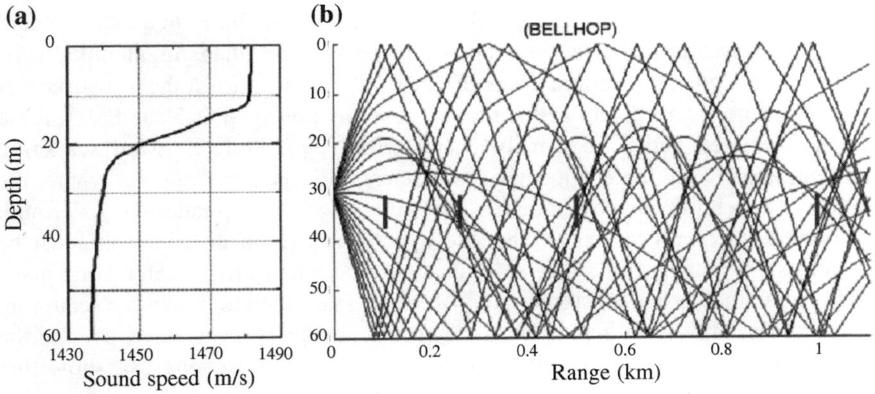

Fig. 2 CAPEx09 propagation environment [10]. **a** Measured sound speed profile; **b** Experimental geometry. Shown are the calculated acoustic rays from the transmitter at a depth of 30 m. Shown to vertical scale are the acoustic receiving arrays at the four ranges (100, 250, 500, 1 km) emphasized in the processing

speed gradient caused results to vary with range in ways that at first seem unexpected: performance at 250 m was better than at 100 m, and performance at 1 km was better than at 500 m. The performance also became more stable as the range increased with less frequent updating of the equalizer required.

The ray trace in Fig. 2 helps explain the range- and depth-dependence observed in communications performance. At 100 m, direct, surface-bounce, and bottom-bounce acoustic paths are observed while the array is positioned near a caustic at 250 m and many strongly refracted direct paths are incident. The strongly refracted paths avoid interaction with the rough surface above resulting in better communications performance. EigenRay calculations at a range of 500 m show that the bottom of the array will see a direct path while the top of the array will not. Evidently, the direct paths are better for communications as a two-channel receiver at the bottom of the array had a bit error rate of only 0.9 % while the top of the array had 2.3 % [10]. At range 1 km, there are no direct paths to any part of the array and each arrival has at least one reflection from the bottom. Rays get trapped in a duct and avoid the rough surface. The result is a stable channel where only infrequent updating of the linear equalizer is required.

4 Performance Prediction Modeling

Kilfoyle and Baggeroer [1] wrote "acoustic propagation models tailored to telemetry applications are sorely needed" and "captur[ing] the time variability of the channel is a necessary component of these models." While the detailed development of such a propagation model is beyond the scope of this short communication, certain key aspects can be outlined.

As argued above, the dynamic rough sea surface is responsible for the most rapid time variability in the acoustic channel. Capturing this variability in a model would seem a daunting task, but several factors make it easier than might initially be thought. For a communications channel, forward scattering from the sea surface is much more important than backscattering. Also, as observed at KauaiEx03, these forward-scattered paths are usually characterized by $P > 1$. Forward scattering configurations with large Rayleigh parameters fall into the regime where the classical Kirchhoff approximation for rough surface scattering is valid. Furthermore, experiments show that the update interval of an equalizer must be much less than the surface wave's dominant period when surface-interacting paths are to be retained as useful signal. Since it is necessary then to model the scattering process over only time scales short compared to the surface wave period, the calculation simplifies considerably. In the regime of interest, one can show that detailed features of the sea surface like capillary waves or those causing the wind direction are of little consequence. One finds that a simple criterion for the usefulness of an acoustic path that bounces off the sea surface is that

$$\frac{(\text{Rayleigh parameter}) \times (\text{total time spread})}{\text{surface wave period}} \ll 1 \qquad (2)$$

In this context, the total time spread includes multi-path spreading, the update interval for the equalizer, and the time window over which the equalizer parameters are estimated. The total time spread is typically much less than the surface wave period implying that Eq. (2) can be satisfied even when the Rayleigh parameter is large compared to one. Detailed calculations of the mean-squared-error for an equalizer's soft demodulation output are in excellent agreement with KauaiEx03 observations at both 1 and 2 km [12].

Acknowledgments The author gratefully acknowledges the colleagues who were instrumental in the success of the above-reviewed experiments and who coauthored the subsequent journal papers: Darrell Jackson, David Dowling, James Ritcey, Warren Fox, and John Flynn (2000 Puget Sound Experiment); Michael Porter, Martin Siderius, Mohsen Badiey, and Aijun Song (KauaiEx03); James Ritcey, Zhongkang Wang, Menglu Xia, and Wen Xu (CAPEx09). The United States Office of Naval Research provided support for the experiments. The author also thanks the PRUAC Organizing Committee for the invitation.

References

1. Kilfoyle, D.B., Baggeroer, A.B.: The state of the art in underwater acoustic telemetry. IEEE J. Oceanic Eng. **25**(1), 4–27 (2000)
2. Rouseff, D., Jackson, D.R., Fox, W.L.J., Jones, C.D., Ritcey, J.A., Dowling, D.R.: Underwater acoustic communications using time reversal. IEEE J. Oceanic Eng. **26**(4), 821–831 (2001)
3. Flynn, J.A., Ritcey, J.A., Fox, W.L.J., Jackson, D.R., Rouseff, D.: Decision-directed passive phase conjugation: equalisation performance in shallow water. Electron. Lett. **37**(25), 1551–1553 (2001)
4. Flynn, J.A., Ritcey, J.A., Rouseff, D., Fox, W.L.J.: Multichannel equalization by decision-directed passive phase conjugation: experimental results. IEEE J. Oceanic Eng. **29** (3), 824–836 (2004)
5. Rouseff, D.: Intersymbol interference in underwater acoustic communications using time-reversal signal processing. J. Acoust. Soc. Am. **117**(2), 780–788 (2005)
6. Porter, M., et al.: The Kauai experiment. In: Porter, M.B., Siderius, M., Kuperman, W.A. (eds.) High Frequency Ocean Acoustics. AIP Proceedings, New York (2004)
7. Rouseff, D., Flynn, J.A., Fox, W.L.J., Ritcey, J.A.: Acoustic communication using time-reversal signal processing: spatial and frequency diversity. In: Porter, M.B., Siderius, M., Kuperman, W.A. (eds.) High Frequency Ocean Acoustics. AIP Proceedings, New York (2004)
8. Rouseff, D., Badiey, M., Song, A.: Effect of reflected and refracted signals on coherent underwater acoustic communication: results from the Kauai experiment (KauaiEx 2003). J. Acoust. Soc. Am. **126**(5), 2359–2366 (2009)
9. Rouseff, D., Wang, Z., Zhou, S., Zhou, L.: The 2009 Cooperative Array Performance Experiment (CAPEx09) comparing vector- and pressure-sensor arrays. In: Proceedings of the European Conference on Underwater Acoustics (ECUA), Istanbul, Turkey, 2010

10. Xia, M., Rouseff, D., Ritcey, J.A., Zou, X., Polprasert, C., Xu, W.: Underwater acoustic communication in a highly refractive environment using SC–FDE. IEEE J. Oceanic Eng. **39** (3), 491–499 (2014)
11. Eckart, C.: The scattering of sound from the sea surface. J. Acoust. Soc. Am. **25**, 566–570 (1953)
12. Rouseff, D., Jackson, D.R.: Simplified formulations for sea-surface scattering for use in modeling equalizer performance in underwater communications. J. Acoust. Soc. Am. **129**(4), 2665 (2011)

Comparisons of Methods for Numerical Internal Wave Simulation in Long-Range Acoustical Propagation

Rex K. Andrew

Abstract The only technique currently known for estimating all the statistical properties of an acoustical field propagating through a randomizing ocean is Monte Carlo simulation. The standard model in deep ocean propagation asserts that the randomization is due to sound speed perturbations caused by the vertical fluid displacements of random internal waves. The Henyey–Reynolds algorithm (Henyey and Reynolds in Numerical simulator of ocean internal waves for long-range acoustics, 2013 [4]) provides a computationally efficient method for generating these displacements for an ocean with range-independent stratification. This method, which is free from the Wentzel–Kramers–Brillouin approximation, uses vertical internal wave modes parameterized only by horizontal wavenumber magnitude, i.e., along only one dimension, as opposed to throughout the two dimensional horizontal wavenumber plane. Results are shown for the standard horizontally isotropic Garrett–Munk spectral model, and compared to the Colosi–Brown algorithm (Colosi and Brown in J Acoust Soc Am 103(4):2232–2235, 1998 [3]). Accurate models of the oceanographic mechanisms causing sound speed randomization will be needed in the estimation of the parameters of ocean mixing processes via inversion of acoustical fluctuation statistics.

Keywords Internal waves · Simulation · Long-range acoustic propagation

1 Background

The most ubiquitous physical mechanism for randomizing the sound speed in the deep ocean is the internal gravity wave field. Models of the internal wave fields evolved throughout the 1970s: the "standard model," known as GM79, was given in 1981 by Munk [1]. Internal waves were first invoked in Monte Carlo acoustical calculations by Dozier and Tappert in 1978 [2]. In 1998, Colosi and Brown [3]

R.K. Andrew (✉)
Applied Physics Laboratory, University of Washington, Seattle, WA 98105, USA
e-mail: rex@snark.apl.washington.edu

L. Zhou et al. (eds.), *Underwater Acoustics and Ocean Dynamics*
DOI 10.1007/978-981-10-2422-1_3

(henceforth CB98) updated the Dozier and Tappert sound speed simulation approach (e.g., to include GM79). Concurrently, Henyey and Reynolds developed an algorithm that has only recently been published [4] (henceforth HR13). Evans also produced a similar algorithm in an unpublished note [5].

The HR13 algorithm is described in the next section, and its performance is compared in the following section to two results from WKB theory. The HR13 algorithm is compared to the CB98 algorithm in the last section, including a summary of the benefits of the HR13 algorithm.

2 Algorithm

The sound speed is represented as $c(\mathrm{r}) = c_0(\mathrm{r}) + \delta c(\mathrm{r}, t)$, where $c_0(\mathrm{r})$ is a "background" field and $\delta c(\mathrm{r}, t)$ is the randomizing component. Here, r represents field position. In the conventional approach, $\delta c = \zeta\left(\frac{\mathrm{d}c}{\mathrm{d}z}\right)_p$, where $\left(\frac{\mathrm{d}c}{\mathrm{d}z}\right)_p$ is the vertical gradient of "potential sound speed" [6], ζ is a stochastic field representing the random vertical displacement of a fluid parcel. HR13 conceptualizes the internal wave field as horizontally isotropic and models, with horizontal plane-wave modes and vertical standing-wave modes.

The vertical modes obey the linearized equation of motion

$$\frac{\mathrm{d}^2 w}{\mathrm{d}z^2} + k^2 \frac{n(z)^2 - \omega^2}{\omega^2 - f^2} w = 0 \tag{1}$$

where f is the local inertial frequency and $n(z)$ is the Brunt–Väisälä frequency profile due to ocean stratification. For a given value of horizontal wavenumber k, Eq. (1) defines a Sturm–Liouville problem. The solution consists of eigenfunctions $w_j(z)$ and associated eigenvalues $\lambda_j(k)$. The eigenvalues can be inverted to provide "eigenfrequencies" $w_j(k)$. HR13 solves Eq. (1) for eigenfunctions and eigenfrequencies by discretizing directly into a matrix eigenvalue problem and using a generalized tridiagonal solver. The group speeds $\frac{\mathrm{d}\omega}{\mathrm{d}k}$ will also be required. These can be calculated directly from the equation $\omega(\lambda)$ yielding

$$\frac{\mathrm{d}w}{\mathrm{d}k} = \frac{\omega^2 - f^2}{\omega k}\left[1 - \left(\omega^2 - f^2\right) \int w^2 \mathrm{d}z\right] \tag{2}$$

using the Hellman–Feynman theorem in Ref. [4] for $\frac{\mathrm{d}\omega}{\mathrm{d}k}$ in terms of the eigenfunctions $w_j(z)$. No approximations are required.

The literature provides two methods for synthesizing homogeneous pseudo-random wavefields. Both utilize an approach specifying the spectral properties of the wavefield. In two spatial dimensions, the first approach, due to Rice [7], incorporates random plane-wave coefficients:

$$\zeta(x,t) = \sqrt{\Delta_{k_x}\Delta_{k_y}}\sum_{k_x}\sum_{k_y}\sqrt{s_{zz}}[q_R\cos(k_x x - \omega t) + q_I\sin(k_x x - \omega t)] \tag{3}$$

where $s_{zz} = s(k_x, k_y)$ is the spectrum of the process, and $q_R \sim N(0,1)$ and $q_I \sim N(0,1)$ independently. The second approach, due to Shinozuka [8], uses randomized phase:

$$\zeta(x,t) = \sqrt{2}\sqrt{\Delta_{k_x}\Delta_{k_y}}\sum_{k_x}\sum_{k_y}\sqrt{s_{zz}}\cos(k_x x - \omega t + \phi) \tag{4}$$

where $\phi = \phi(k_x, k_y)$ is a uniform variate in $[0, 2\pi]$. The statistical distributions of these two methods differ: Katafygiotis et al. [9] discuss this issue in detail. For our purposes, note that each method accumulates contributions over the (k_x, k_y) wavenumber plane. The two methods described above raster over a uniform grid in (k_x, k_y) space. HR13 chooses an alternate scheme, rastering over k_x and $K = \sqrt{k_x^2 + k_y^2}$. This choice incurs variable-sized differential elements shown in Fig. 1.

Given independent sampling in k, HR13 utilizes logarithmic spacing, $k_m = k_0 r^m$, $m = 0, 1, 2, 3, \ldots$. Here, k_0 is the minimum wavenumber (maximum spatial scale), and r defines the number of samples per decade. HR13 uses $r = 10^{1/25}$, i.e., 25

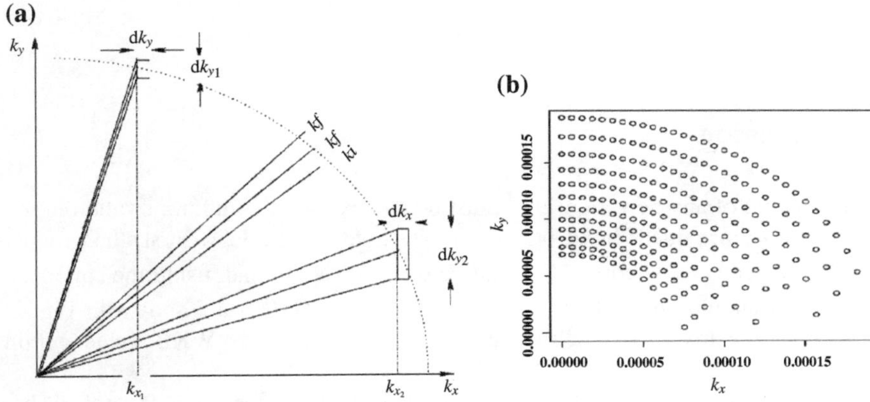

Fig. 1 Tiling strategy for the plane. **a** Example of the differential areal tiling based on the sampling scheme. The example presents only the sampling in the first quadrant of the plane. **b** Example of the logarithmic sampling in the first quadrant of the plane. Only the first 12 values of horizontal wavenumber k are shown, corresponding to $m = 0, 1, 2, \ldots, 11$. The logarithmic spacing in k is already evident. The gridding is uniform in k_x. For a single mode number j, only 12 modes need be computed to support this constellation of sample points in space

samples per decade. A representative diagram of this sampling scheme is shown in Fig. 1.

The displacement spectrum is derived from GM79 by removing the WKB approximations in Ref. [4] and is

$$s_{zz}(\omega,j) = \frac{2E_0B(N_0B)^2}{\pi M\left(j^2+j_*^2\right)}\frac{f}{\omega^3}\sqrt{\omega^2-f^2} \tag{5}$$

which is defined in terms of mode number j and frequency ω. This needs to be converted to a function of mode number j and horizontal wavenumber k, as follows:

$$s_{zz}(\omega,j)\mathrm{d}\omega = s_{zz}(\omega,j)\mathrm{d}\omega\frac{\mathrm{d}\theta}{2\pi} = \frac{s_{zz}(\omega_j(k),j)}{2\pi k}\frac{\mathrm{d}\omega}{\mathrm{d}k}\mathrm{d}k_x\mathrm{d}k_y \tag{6}$$

where $M = \sum_{j=1}^{JMAX}(j^2+j_*^2)^{-1}$. Then, using

$$\left\langle |G_+|^2\right\rangle + \left\langle |G_-|^2\right\rangle = \frac{s_{zz}(\omega_j(k),j)}{2\pi k}\frac{\mathrm{d}\omega}{\mathrm{d}k}\mathrm{d}k_x\mathrm{d}k_y \tag{7}$$

where G_+ and G_- are complex Gaussian variants, the final expression for the slice $\zeta(x,z;t)$ is

$$\zeta(x,z;t) = F^{-1}\left\{\sum_{j=1}^{JMAX}\sum_{k_y}\left[G_-\mathrm{e}^{-i\omega t}+G_+\mathrm{e}^{+i\omega t}\right]W_j(z)\right\} \tag{8}$$

using an inverse Fourier transform in k_x.

3 Validation

A key feature of any new algorithm must be an auxiliary definition of valid output. Two tests are provided. Input parameters are given in Table 1. First, the product $n(z)\langle\zeta^2\rangle$ under WKB theory [1] should be depth invariant and, using the canonical test values, should equal 0.28. Five realizations of $n(z)\langle\zeta^2\rangle$ are shown in Fig. 2: throughout the main thermocline, this algorithm attains the WKB value within sampling error.

The second test compares the horizontal wavenumber spectrum of vertical displacements. This is essentially a "towed" spectrum. The WKB result generally follows a k_x^{-2} dependence at high wavenumber. A sample spectrum estimated from a single horizontal series and the corresponding theoretical expression are shown in Fig. 2. An approximate high-wavenumber WKB solution derived by Levine et al.

Table 1 Parameters used in the algorithm, with values used in the simulations

Description	Value	Parameter
Maximum x range	1×10^6 m	
# Grid points in x direction	8192	
# Grid points in z direction	512	
Total modes	80	*JMAX*
# Horizontal wavenumber grid points	100	
Maximum horizontal mode scale	1×10^5 m	$= 2\pi/k_0$
Minimum horizontal mode scale	1×10^3 m	
Brunt-Väisälä profile	Munk canonical profile	
Brunt-Väisälä profile scale	3.0 cph	N_0
Brunt-Väisälä profile depth scale	1300 m	B
GM strength	0.0832 m	$E_0 B$
Horizontal wavenumber gridding base	$10^{1/25}$	
Grid steps per decade	25	
Latitude	20°	
Modal bandwidth	3.0	j_*

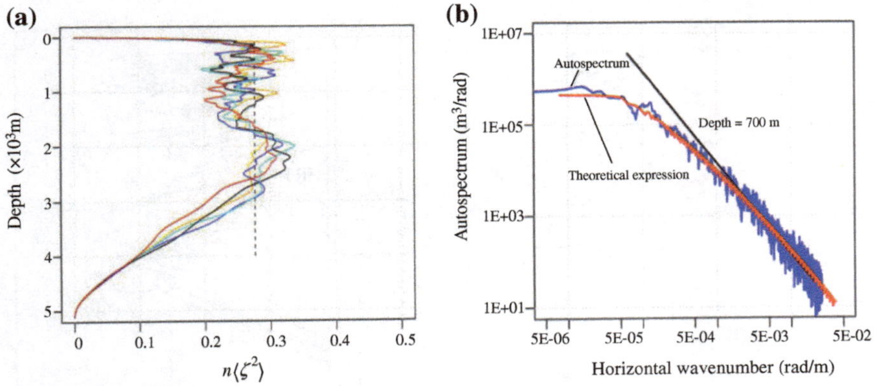

Fig. 2 Validation tests. **a** Simulation realizations of the product $n(z)\langle\zeta^2\rangle$ using the HR13 model and random number generator seeds of −1, −2, −3, −4, −5. The vertical line is the WKB result at 0.28. **b** Comparisons of horizontal "towed" spectra, at the depth of 700 m. A single realization is used. The sample variance was 92.74. The autospectrum was estimated using the multitaper method with 8 tapers and a time-bandwidth product of 4. The estimated integral of the sample autospectrum was 93.77. The discretization of the sampling in wavenumber space is evident as a "stair-step" appearance in the theoretical curve at higher wavenumbers. The predicted variance based on an estimated integral of this curve is 83.97. The Levine approximate expression is shown with a *solid black line*

[10] is shown as well. The sample spectrum has excellent agreement at high wavenumbers, indicating that the algorithm generates simulated processes with the correct spatial spectrum slope at high wavenumbers.

4 Comparisons

CB98 provides a Rice method algorithm Eq. (3) and a Shinozuka method algorithm Eq. (4). They also provide simplifications for single (x, z) slice simulations. They use WKB stretching to substitute sinusoids for exact mode shapes. Sample spacing in the (k_x, k_y) wavenumber plane is uniform. The number of the first quadrant wavenumber integrand sample points is roughly 1000×100 for HR13 and 1000×1000 for CB98. The number of modes for HR13 is $100 \times JMAX$ but only *JMAX* for CB98 because their modes are independent of horizontal wavenumber. CB98 specifies the domain of integration as $k_{\min} \leq |k_x|,\ k_y \leq |k_{\max}|$. To compare with HR13, we use instead the domain $k_{\min} \leq \sqrt{k_x^2 + k_y^2} \leq k_{\max}$. The results for the product $n(z)\langle\zeta^2\rangle$ are shown in Fig. 3. Model parameters were taken from Table 1. HR13 simulations use 4 decades of wavenumbers: CB98 simulations use 2. Qualitatively, the results are similar; quantitatively, the HR13 method includes a correction for the "interior" region $\sqrt{k_x^2 + k_y^2} \leq k_{\min}$ and thereby achieves the WKB result throughout the main thermocline and the deep sound channel.

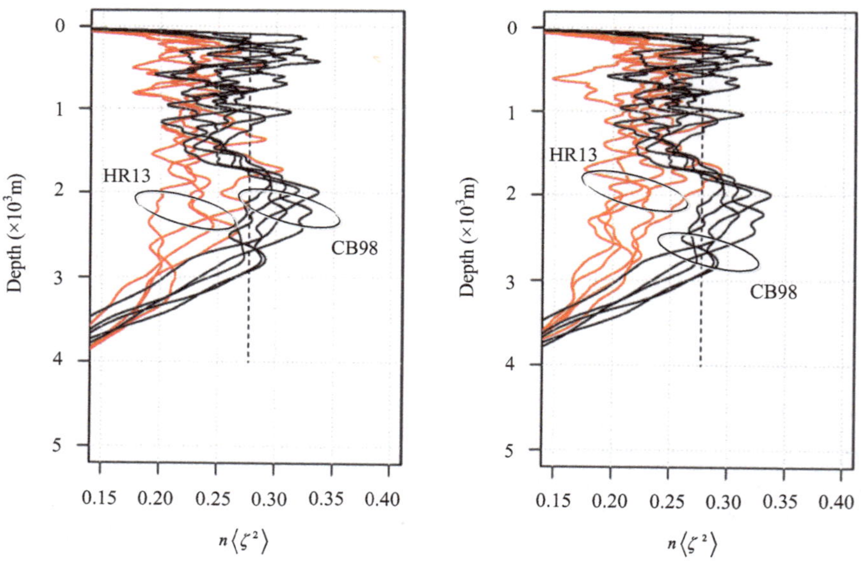

Fig. 3 Comparison of $n(z)\langle\zeta^2\rangle$ for HR13 (Rice) versus CB98 equations 9 (*left*, Rice) and 19 (*right*, Shinozuka)

Acknowledgments Investigation was sponsored by the Office of Naval Research code 322OA Ocean Acoustics.

References

1. Munk, W.H.: A survey of internal waves and small scales processes. In: Warren, B.A., Wunsch, C. (eds.) Evolution of Physical Oceanography. MIT Press, Cambridge, MA (1981)
2. Dozier, L.B., Tappert, F.D.: Statistics of normal mode amplitudes in a random ocean. I. Theory. J. Acoust. Soc. Am. **63**, 353–365 (1978)
3. Colosi, J.A., Brown, M.G.: Efficient numerical simulation of stochastic internal-wave-induced sound-speed perturbation fields. J. Acoust. Soc. Am. **103**(4), 2232–2235 (1998)
4. Henyey, F.S., Reynolds, S.A.: Numerical simulator of ocean internal waves for long-range acoustics. Technical Memorandum APL-UW TM 1-13. Applied Physics Laboratory, Seattle Washington, 2013
5. Evans, R.B.: Calculation of internal-wave eigen-frequencies and modes; displacement and sound speed realizations. Unpublished technical note, URL http://oalib.hlsresearch.com/Other/wave/wave.doc (2000)
6. Munk, W.H., Zachariasen, F.: Sound propagation through a fluctuating stratified ocean: theory and observation. J. Acoust. Soc. Am. **59**(4), 818–838 (1976)
7. Rice, S.O.: Mathematical analysis of random noise. In: Wax, N. (ed.) Selected Papers on Noise and Stochastic Processes, pp. 133–294 (1954)
8. Shinozuka, M.: Stochastic fields and their digital simulation. In: Schuëller, G.I., Shinzuka, M. (eds.) Stochastic Methods in Structural Dynamics. Martinus Nijhoff, Dordrecht (1987)
9. Katafygiotis, L.S., Zerva, A., Malyarenko, A.A.: Simulation of homogeneous and partially isotropic random fields. J. Eng. Mech., 1180–1189 (1999)
10. Levine, M.D., Irish, J.D., Ewart, T.E., Reynolds, S.A.: Simultaneous spatial and temporal measurements of the internal wave field during MATE. J. Geophys. Res. **91**(C8), 9709–9719 (1986)

Acoustic Data Assimilation: Concepts and Examples

Liling Jin, Wen Xu, Jiamin Huang and Jianlong Li

Abstract The ocean is a complicated hydro-dynamic system, showing various interdisciplinary processes of multiple interactive scales. Such processes as internal waves, eddies, and fronts make the ocean acoustic environment drastic changes in time and space. Understanding and modeling physical-acoustical processes are essential for ocean acoustic applications as well as ocean field prediction and parameter estimation. The acoustic data assimilation (ADA), which melds instant observed data of different natures and various physical models, has recently been developed as a new technique for forecasting both ocean and acoustic fields. The general framework of ADA includes three major parts: (1) an observational network for data measurements; (2) a suite of interdisciplinary ocean physics models and the sound propagation model; and (3) data assimilation schemes. Hence it is expected to provide ocean acoustic predictions with higher resolution and better accuracy, compared to those only using the observation data or the physics model. In this paper, we discuss the concepts and framework required to develop the coupled physics-acoustical data assimilation schemes, and review some typical ADA systems and their filed-testing results. It is shown that both the theory and related experiments are advancing steadily.

Keywords Acoustic data assimilation · Ocean acoustic predictions · Parameter estimation · Physical-acoustical processes

L. Jin (✉) · W. Xu · J. Huang · J. Li
College of Information Science and Electronic Engineering, Zhejiang University, Hangzhou 310027, China
e-mail: wxu@zju.edu.cn

J. Li
e-mail: jlli@zju.edu.cn

L. Zhou et al. (eds.), *Underwater Acoustics and Ocean Dynamics*
DOI 10.1007/978-981-10-2422-1_4

1 Introduction

In the ocean, interdisciplinary processes, involving physical, biological, chemical, sedimentological, acoustical, and optical processes, are now known to occur on multiple interactive scales in space and time. These processes cover a wide range of space scales from 1 mm to 10,000 km, and of time scales from 1 s to 100 years and more [1]. Understanding and modeling these processes are essential for ocean field prediction and parameter estimation. However, conventional oceanographic measurements cannot carry out a large area and long-term observation with sufficient temporal and spatial sampling, especially for the sub-mesoscales processes [2]. Recent research on coupled oceanographic and acoustic data assimilation (ADA) is concerned with melding instant observed data of different natures and various physical models. A large number of simultaneous physical, biological, and acoustical measurements and physical models are combined by ADA to study the physical-acoustical-biological processes. Importantly, efforts of ocean field prediction and parameter estimation are enhanced by integrating acoustical measurements. ADA is becoming a common technique for forecasting both ocean and acoustic fields.

In this paper, we first discuss the concepts and framework required to develop coupled physics-acoustical data assimilation schemes, and then review some typical ADA systems and their field-tested results. In the end, we draw some conclusions and introduce a number of future trends based on our understanding toward the problem.

2 Concepts of Acoustic Data Assimilation

ADA melds estimates of data and models for field and parameters with three major parts: (1) an observational network for data measurements; (2) a suite of interdisciplinary ocean physics models and the sound propagation model; and (3) data assimilation schemes. The diagram of the ADA is shown in Fig. 1, which illustrates

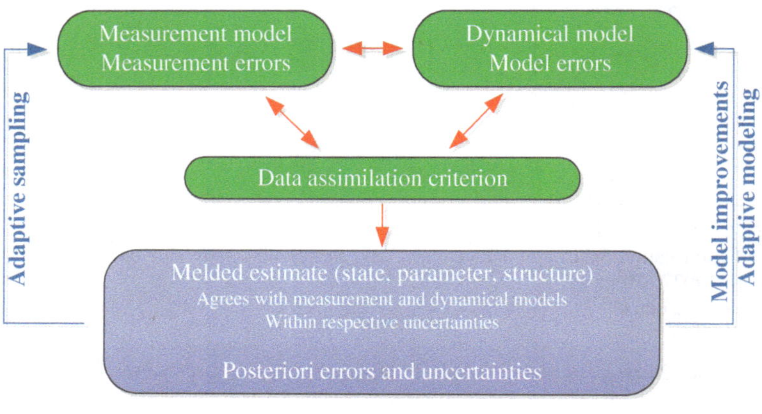

Fig. 1 Diagram of the acoustic data assimilation (reprinted from Ref. [3])

the system components and their interactions, including two feedbacks [3]. Data measurements are linked to the state variables and dynamical model via measurement models. Dynamical models represent the interdisciplinary ocean physics models and the sound propagation model. Then, based on the data assimilation criterion, data and dynamics are melded with inverse weighting related to their relative errors. Finally, the melded estimates are used to improve the model predictions and optimize the resource deployment of an observational network.

2.1 Data

Presently, advanced measurement techniques present powerful opportunities for the acquisition of extensive data sets. The platforms can be used to organize an observational network, such as ships, underwater vehicles, etc. The data required by the ADA system include the following data of different characteristics:

(1) in situ dynamical data, e.g., profiles of temperature, salinity, and current velocities from XBTs, CTDs, ADCPs;
(2) remotely sensed data, mostly satellite data;
(3) acoustic data, including sound pressure, travel time, and transmission loss.

2.2 Models

Ocean dynamics are governed by Newton's equations of motion expressed as the Navier–Stokes equations. Many ocean models have been developed to numerically solve the governing equations (which are also referred to as the primitive equations), such as Harvard Ocean Prediction System (HOPS) [4], Princeton Ocean Model (POM) [5], Regional Ocean Model System (ROMS) [6], and The Unstructured Grid Finite Volume Coastal Ocean Model (FVCOM) [7]. For computational economy, the primitive equations are often solved under the hydrostatic and Boussinesq assumptions. However, for some special dynamical processes, the assumptions may become inappropriate. Taking internal waves as an example, the governing equations should be replaced by non-hydrostatic primitive equations or dedicated internal wave equations. Although the ocean models have become powerful tools for marine environment simulations, the errors caused by numerical calculations and the uncertainty of the boundary conditions and atmospheric forcing are still inevitable. Therefore, assimilation methods should be employed to calibrate the ocean models using the measurement data.

2.3 Assimilation Methodology

A number of methods are available for coupled oceanographic and ADA, which can be divided into four classes [8]

(1) estimation theory, such as the Kalman smoother and optimal interpolation;
(2) control theory, such as the adjoint method;
(3) direct minimization methods, such as the steepest descent, conjugate gradient, simulated annealing, and genetic algorithms;
(4) stochastic and hybrid methods, such as the Error Subspace Statistical Estimation (ESSE).

3 Current Status and Main Results

In this section, we introduce a few typical ADA systems and their field-tested results.

3.1 MIT Ocean Acoustic Tomography Framework

A general framework of ADA was first proposed at MIT to reduce the environmental uncertainty and estimation the true environment parameters [9]. In this approach, a variety of information sources are exploited, including a direct local sound speed measurement model, an oceanographic dynamic model of the sound speed field, and a full field acoustic propagation model as well as other various measurements. These four sources of information are described as

$$\tilde{c} = Mc + c_n, \quad c_n \sim N(0, R_c) \tag{1}$$

$$A(c) = v, \quad v \sim N(0, Q_a) \tag{2}$$

$$B(p, c) = w, \quad w \sim N(0, Q_b) \tag{3}$$

$$\tilde{p} = Lp + p_n, \quad p_n \sim N\left(0, R_p\right) \tag{4}$$

where Eq. (1) denotes the local sound speed measurement model and c_n is the measurement error of the local sound speed, Eq. (2) denotes the oceanographic dynamic model and v is the error of the model, Eq. (3) denotes the full field acoustic propagation model and w is the modeling error, Eq. (4) denotes the acoustic pressure measurement model and p_n is the measurement noise. $\tilde{c}$ and $\tilde{p}$ denote the local sound speed measurements and the local acoustic pressure measurements; c and p denote the true sound speed field and acoustic pressure field; M and L are the mapping matrices; and R_c, Q_a, Q_b, and R_p denote the covariance matrices. All the additive error terms are modeled as a Gaussian random distribution. Then, the paper introduces a variational method to solve this inverse problem, through minimizing the misfit (also called the cost function) between the measurements and their predictions calculated from the available models

$$\hat{c} = \arg\min_{p,c}\{J(p,c)\} \tag{5}$$

where the cost function J can be described as

$$J(p,c) = c_n^{\mathrm{H}} R_c^{-1} c_n + v^{\mathrm{H}} Q_a^{-1} v_n + x^{\mathrm{H}} Q_b^{-1} x + p_n^{\mathrm{H}} R_p^{-1} p_n \tag{6}$$

This article is the first time toward the assimilation of acoustic data into ocean models, which is referred to as an ADA. However, Ref. [9] just verifies the feasibility of the proposed framework, and the oceanographic dynamic model was not included. Following the approach in Ref. [9], a modified GM internal wave model for the specific shallow-water environment is included as an oceanographic dynamical model in the framework of ADA [10]. In the numerical examples, the true environment parameters can be estimated more precisely with the iterative implementation of the inversion process. Analyzing the simulation results, we find that the inversion performance is quite well, which further verify the validity of the ADA approach in ocean environmental inversion.

3.2 *Harvard Ocean Prediction System*

The interdisciplinary ocean observing and prediction system developed at Harvard and presently in scientific and operational use is HOPS [4]. HOPS is an integrated system of data assimilation schemes. The PE physical dynamical model is utilized in HOPS, as an oceanographic dynamic model. Three coordinate transformations are available for the vertical coordinate, including sigma, hybrid and multiple sigma coordinate transformations. In the horizontal dimension, multiple two-way nests are included. The entire system includes a data analysis and management module, data assimilation, dynamical consistent model initialization, and model-driven adaptive sampling with feedbacks.

During July and August of 1996, the US Office of Naval Research (ONR) Shelfbreak PRIMER experiment was carried out in the Mid-Atlantic Bight south of New England. The main objective was to study the effects of oceanographic changes on the sound propagation. We found that the posterior TL is substantially closer to the true TL than the prior, and the estimation performance of the assimilation of the TL and sound speed data is better than the assimilation of the TL data alone [11].

3.3 *Focused Acoustic Forecasting-05*

For limited environment measurements and computational capabilities, oceanic acoustic predictions cannot be provided in high resolution and with enough

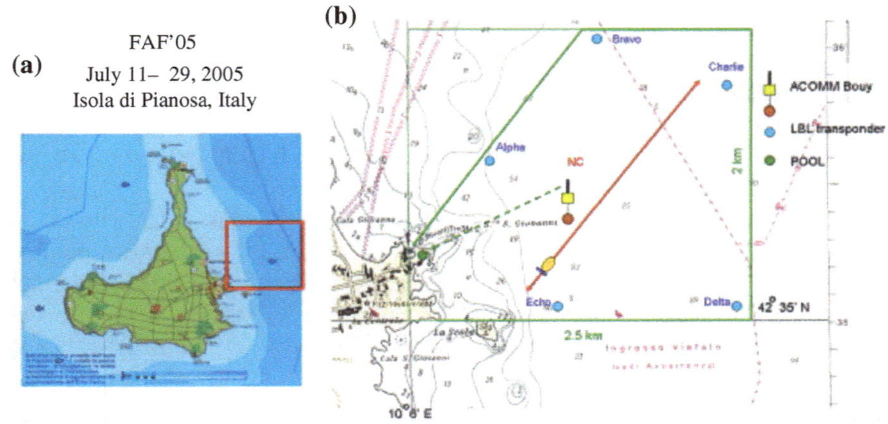

Fig. 2 **a** Geographic location of the FAF05 experiment; **b** General schematic of the FAF'05 experiment [12]

accuracy. Adaptive Rapid Environmental Assessment (AREA) is an adaptive acoustics-environment sampling approach being developed. By assimilating in situ measurements sampled adaptively and optimally, AREA can dramatically improve the estimation performance. In 2005, the Focused Acoustic Forecasting-05 (FAF'05) exercise was held to illustrate these concepts. FAF'05 was carried out in the northern Tyrrhenian sea, on the eastern side of the Corsican channel (Fig. 2). The Autonomous Underwater Vehicle (AUV) domain covers an area of about 2.5 km × 2.5 km. The main realization for the FAF'05 exercise simulations can be summarized in three steps: (1) HOPS/ESSE was used to predict ocean environmental daily; (2) acoustic forecasts were computed by various scenario predictions of sound speed using a sound propagation model; (3) sound speed and acoustic forecasts were input into an adaptive sampling, and an optimal prior AUV path was forecasted based on optimization algorithm, via which the corresponding uncertainties of the predicted acoustic field were reduced. The results of the FAF'05 simulations clearly demonstrated the validity of the AREA methodology.

3.4 Battlespace Preparation 2007

The real-time coupling data assimilation system of a nested ocean model and an acoustic propagation model were carried out at sea for the first time, as a part of the Battlespace Preparation 2007 (BP07). The operational region of BP07 was southeast of the island of Elba (Italy) in the Tyrrhenian Basin, east of Corsica and Sardinia. During BP07, a new real-time working process chain was developed prior to the exercise, as shown in Fig. 3. It consists of the following parts: (1) in situ ocean data collected based on oceanographic forecasts; (2) the in situ ocean data

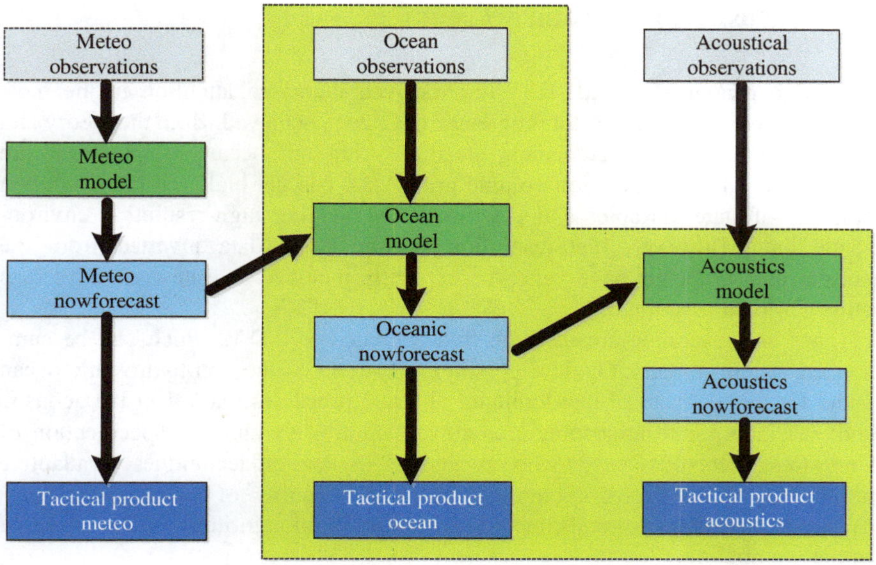

Fig. 3 Working process chain of the data-driven oceanographic acoustic modeling system [13]

and external meteo-data assimilated into an oceanographic modeling system to provide oceanographic forecasts; (3) sound speed forecasts then used as inputs to the acoustic model for acoustic propagation forecasts. Some of the results are given in Ref. [13]. We can find that the ocean environment variability can highly influence sound propagation properties and the probability of detection. The coupling of at-sea acoustic modeling to real-time ocean forecasting is essential.

3.5 Quantifying, Predicting, and Exploiting Uncertainty

The 2008 Quantifying, Predicting, and Exploiting Uncertainty (QPE2008) pilot experiment was carried out on the continental shelf and slope of northeast Taiwan, China [14]. The QPE2008 was designed to (1) develop real-time methods to forecast the ocean, the seabed, and the acoustics; (2) compare predictions from such coupled systems to in situ data; (3) study and quantify the influences of the ocean and geo-acoustic uncertainties on the coupled predictions. In Ref. [14], the authors quantified the dynamic mechanisms and the sensitivity with acoustic transmission data collected from the QPE2008. From that paper, one can obtain that the sediment properties lead to larger but isotropic variations on the shelf and smaller but anisotropic variations over the shelfbreak, and internal tides on the shelf lead to the largest TL sensitivity. In general, those results are useful for improving the real-time modeling of coupled physics-acoustical data assimilation and optimum sampling network.

4 Conclusions and Future Trends

Coupled oceanographic and ADA has received increased attention in the most recent ten years, and significant progresses have been achieved. Both the theory and related experiments are advancing steadily. Although ocean computations are restricted to a lower resolution volume grid, ADA can get high vertical resolution estimates of state variables and parameters by melding high-resolution environmental data. However, high-resolution environmental data inverted from the acoustic measurements have yet to be explicitly included as a new effective data source for data assimilation.

There are several future research trends related to ADA, which can be summarized in four aspects: (1) develop solving systems of the non-hydrostatic ocean model for the estimate of the dominant physical processes that affect the acoustic field; (2) incorporate atmospheric adaptive sampling to improve specification of wind stress, heat flux, etc., to ocean models; (3) exploit techniques of adaptive sampling, such as network building and path planning of moving platforms; (4) design algorithms for real-time ADA, such as rapid algorithms and methods of parallel computing.

Acknowledgments This work was supported by the National High Technology Research and Development Program of China (grant No. 2012AA090901), the Fundamental Research Funds for the Central Universities (grant Nos. 2011XZZX003 and 2013XZZX011), and the National Natural Science Foundation of China (grant No. 61171147). The authors want to thank all the researchers in this area for their outstanding work.

References

1. Dickey, T.D.: Technology and related developments for interdisciplinary global studies. Sea Technology, 47–53 (1993)
2. Dickey, T.: Emerging ocean observations for interdisciplinary data assimilation systems. J. Mar. Syst. **40**(1), 5–48 (2003)
3. Robinson, A.R., Lermusiaux, P.F.J.: Prediction systems with data assimilation for coupled ocean science and ocean acoustics. In: The Sixth International Conference on Theoretical and Computational Acoustics (ICTCA), Honolulu, HI, Aug 2003
4. Robinson, A.R.: Forecasting and simulating coastal ocean processes and variabilities with the Harvard Ocean Prediction System. In: Mooers, C.N.K (ed.) Coastal Ocean Prediction. AGU Coastal and Estuarine Studies Series, American Geophysical Union, pp. 77–100 (1999)
5. Mellor, G.L.: Users guide for a three dimensional, primitive equation, numerical ocean model. Program in Atmospheric and Oceanic Sciences, Princeton University, 1998
6. Shchepetkin, A.F., McWilliams, J.C.: The regional oceanic modeling system (ROMS): a split-explicit, free-surface, topography following-coordinate oceanic model. Ocean Model. **9**, 347–404 (2005)
7. Chen, C.S., Liu, H.D., Beardsley, R.C.: An unstructured, finite-volume, three-dimensional, primitive equation ocean model: application to coastal ocean and estuaries. J. Atmos. Oceanic Technol. **20**, 159–186 (2003)

8. Robinson, A.R., Lermusiaux, P.F.J., Sloan, N.Q.: Data assimilation. In: Brink, K.H., Robinson, A.R. (eds.) THE SEA: The Global Coastal Ocean. Processes and Methods, vol 10, pp. 541–594. Wiley, New York (1998)
9. Elisseeff, P., Schmidt, H., Xu, W.: Ocean acoustic tomography as a data assimilation problem. IEEE J. Oceanic Eng. **27**(2), 275–282 (2002)
10. Li, J.L., Jin, L.J., Xu, W.: Inversion of internal wave-perturbed sound speed field via acoustic data assimilation. IEEE J. Oceanic Eng. **39**(3), 407–418 (2013)
11. Lermusiaux, P.F.J., Chiu, C.S.: Four-dimensional data assimilation for coupled physical-acoustical fields. In: Pace, N.G., Jensen, F.B. (eds.) Acoustic Variability, pp. 417–424. Kluwer Academic Press, Saclantcen (2002)
12. Wang, D., Lermusiaux, P.F.J., Haley, P.J., Eickstedt, D., Leslie, W.G., Schmidt, H.: Acoustically focused adaptive sampling and on-board routing for marine rapid environmental assessment. J. Mar. Syst. **78**, s393–s407 (2009)
13. Lam, F.P.A., Haley Jr., P.J., Janmaat, J., Lermusiaux, P.F.J., Leslie, W.G., Schouten, M.W., te Raa, L.A., Rixen, M.: At-sea real-time coupled four-dimensional oceanographic and acoustic forecasts during Battlespace Preparation 2007. J. Mar. Syst. **78**, s306–s320 (2009)
14. Lermusiaux, P.F.J., Xu, J.S., Chen, C.F., Jen, S., Chiu, L.Y., Yang, Y.J.: Coupled ocean-acoustic prediction of transmission loss in a continental shelfbreak region: predictive skill, uncertainty quantification, and dynamical sensitivities. IEEE J. Oceanic Eng. **35**(4), 895–916 (2010)

The Preliminary Results of a Single-Hydrophone Geoacoustic Inversion for Data Collected at the Sea of Japan

Pavel S. Petrov

Abstract The geoacoustic inversion method based on the use of warping operators is applied to data collected during experiments in the autumn of 2012 at the Sea of Japan. In these experiments, broadband sound pulses emitted by a compact transducer were recorded by an acoustic station equipped with a digital radio buoy. The recorded signals are used for the estimation of acoustic properties of bottom sediments. The warping transformation is applied to the signal in order to perform mode filtration on the signal's spectrogram. This filtration allows us to estimate group velocities of different modes which are then used as the data for inversion. The results of the inversion are discussed.

Keywords Geoacoustic inversion · Warping transform · Normal modes · Dispersion curves

1 Introduction and Experiments Description

In this paper, we apply recently developed warping-based methods of single-hydrophone geoacoustic inversion [1, 2] to the data collected during experiments in the autumn of 2012 conducted on the shelf of the Sea of Japan in Posyet Bay near the "Cape Schultz" marine research station (Fig. 1a). During these experiments, a compact sound transducer (Fig. 1b) was used to emit the short pulses which were recorded by a hydrophone attached to a digital radio buoy.

The digital radio buoy (DRB) attached to the hydrophone near the sea bottom was deployed in Vityaz Bay, while the transducer (S) was submerged into the seawater to a depth of 10 m from a research vessel. The respective position of the

P.S. Petrov (✉)
V.I.Il'ichev Pacific Oceanological Institute of the Russian Academy of Sciences, Vladivostok 690041, Russia
e-mail: petrov@poi.dvo.ru

P.S. Petrov
School of Natural Sciences, Far Eastern Federal University, Vladivostok 690950, Russia

L. Zhou et al. (eds.), *Underwater Acoustics and Ocean Dynamics*
DOI 10.1007/978-981-10-2422-1_5

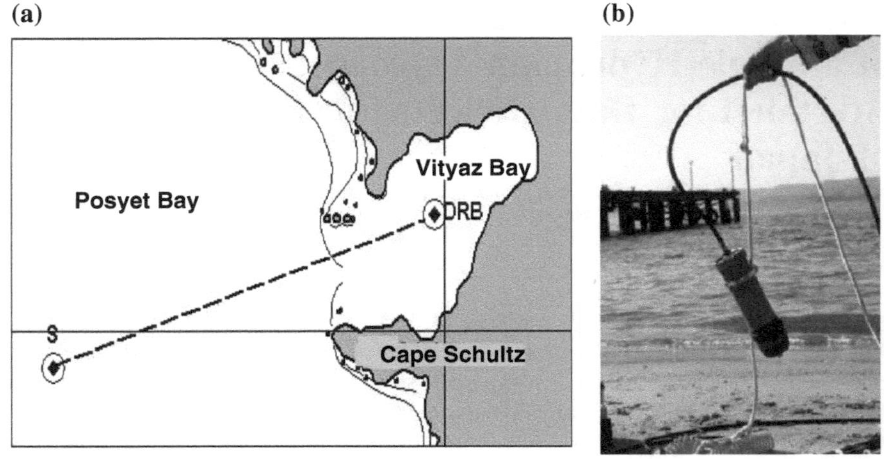

Fig. 1 Scheme of the experiments conducted in the autumn of 2012 in Posyet Bay (**a**) with source (*S*) and receiver (*DRB*) positions indicated and the compact pulse sound transducer photo (**b**)

vessel and the buoy is shown in Fig. 1. The length of the S-DRB track is 5.9 km. The bottom depth along the track S-DRB varies within the interval from 27 to 33 m. Note that the algorithm of the geoacoustic inversion [1, 2] employed in this paper assumes that the bottom depth is constant along the track. For the tracks where depth is slightly range dependent, the algorithm may recover only some estimates of the density and sound speed at the bottom but the degree of uncertainty is still not well understood.

A time series of the pulse signal recorded by the DRB which is used in the subsequent sections is shown in Fig. 2a. The spectrum of this pulse has a very sharp peak at the frequency of 73.5 Hz (Fig. 2b).

The success of the geoacoustic inversion method based on the group velocities estimation from modal and intramode dispersion information [1, 2] strongly depends on the "modal diversity" of the signal. This implies that the richer modal structure of the signal leads to higher precision of the environment reconstruction. By "rich modal structure" we mean:

(1) the diversity of the signal's frequency content (which is usually achieved by the uniformity of a frequency distribution of the signal's spectrum; that is to say, the signal spectrum has to be roughly flat at certain frequency intervals);
(2) the diversity of the modal content of the acoustic field at the frequencies where the energy of the signal is concentrated (i.e., the frequencies must be chosen so that in a given waveguide, sufficiently many propagating waterborne modes [3] were excited).

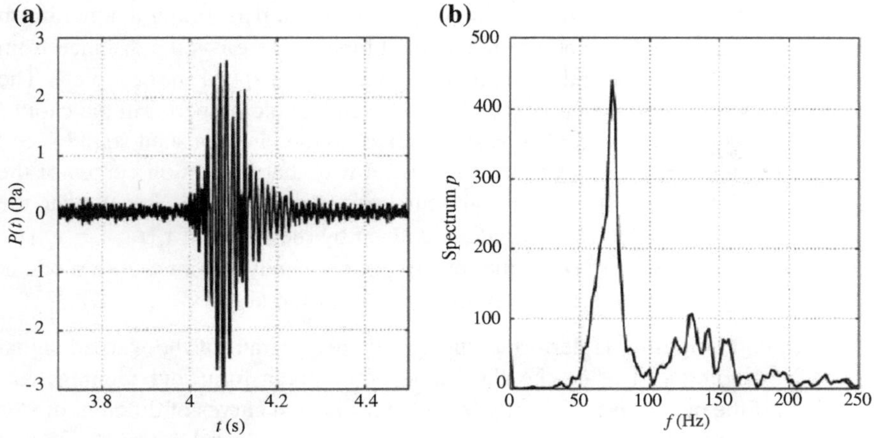

Fig. 2 The signal emitted by the source (*S*) and recorded by the DRB (**a**), 5.9 km away from the source and its spectrum (**b**). Note the sharp peak in the spectrum at the frequency of 73.5 Hz

Sometimes, however, it is impossible to employ the sources producing the signals of appropriate "modal diversity" and one has to extract as much information on the propagation environment as possible from a signal with a relatively poor modal structure. In this paper, we investigate the possibilities of media parameters extraction from the dispersion information under very unfavorable conditions. Indeed, we use a source with a relatively narrowband spectrum (which presents very limited intramode dispersion information) and the source dominant frequency is such that only 1−2 waterborne modes exist for the given bottom depth (the exact number depends on the sound speed and density in the bottom). Our results show that even under these conditions some waveguide parameters can be reliably estimated.

2 Warping-Based Inversion Algorithm Outline

In this section, we briefly recall the inversion scheme developed in [1, 2] and successfully applied in [1] and highlight the results from some other papers.

In the warping-based single-hydrophone inversion algorithm, the following sequence of steps is performed:

(1) The time series $p(t)$ of the original signal recorded by the receiver undergoes the so-called time-warping transform:

$$W_h^t p(t_w) = \sqrt{h'(t_w)}\,p(h(t_w)) \tag{1}$$

where $t = h(t_w) = \sqrt{t_w^2 + r^2/c_{\text{ref}}^2}$ is the warping function performing a nonuniform resampling of the original time t to the warped time t_w (here r is the distance from the source to the receiver and c_{ref} is a "reference" sound speed in the water). The warping transform Eq. (1) enjoys the following remarkable property: in the case of an ideal waveguide with a rigid (non-penetrable) bottom and constant sound speed c_{ref} in water, the signal is transformed in such a way that dispersion curves of the warped signal $W_h^t p(t_w)$ are merely straight horizontal lines [4]. Recall that the dispersion curve [3, 4] of a waveguide is defined by the function $\tau_i(f) = r/v_i^{\text{gr}}(f)$, where f is the frequency, $v_i^{\text{gr}}(f)$ is the modal group velocity [3] for the ith mode at the frequency f and $\tau_i(f)$ is the delay time of the ith mode.

(2) The mode filtration is performed using the spectrogram of the warped signal $W_h^t p(t_w)$ produced using the short-time Fourier transform and running window. If the range r is sufficiently large the dispersion curves of different modes will become separable from each other. They can be filtered out assuming that they are horizontally elongated entities.
(3) The filtered dispersion curves can be transformed back to the "original" time domain and turned into thin lines using reassignment technics. The inverse warping is performed using the same formula Eq. (1) but using the function $t_w \equiv h^{-1}(t) = \sqrt{t^2 - r^2/c_{\text{ref}}^2}$ instead of the warping function $h(t_w)$.
(4) In the original time domain, we obtain the dispersion curves in the form $\tau_i = \tau_i(f)$ and these mode delays can be used as the data for geoacoustic inversion.
(5) To perform the inversion, we construct a cost function:

$$F(A, \mathrm{d}\tau) = \sum_{i=1}^{N_m} \sum_{j=1}^{N_f} \left| \frac{r}{v_i^{\text{gr}}(f_j, A)} - \tau_{\exp}(f_j, i) + \mathrm{d}\tau \right| \tag{2}$$

for a parameter set $A = (c_b, \rho_b, H)$ and time shift $\mathrm{d}\tau$ (c_b, ρ_b, H stand for the sound speed in the bottom, bottom density and the depth, respectively; f_j, $j = 1, \ldots, N_f$ is the frequency grid—the set of frequencies where the model and experiment delays are compared and N_m—number of the modes). The cost function has to be minimized using any suitable optimization algorithm. In this paper, we use a simple exhaustive search optimization (since the search space has four dimensions, the optimization takes only 3 h on a midrange laptop).

3 Application of the Inversion Scheme to Our Data

Now we can proceed to the application of the inversion method outlined in the previous section to the signals recorded in our experiments.

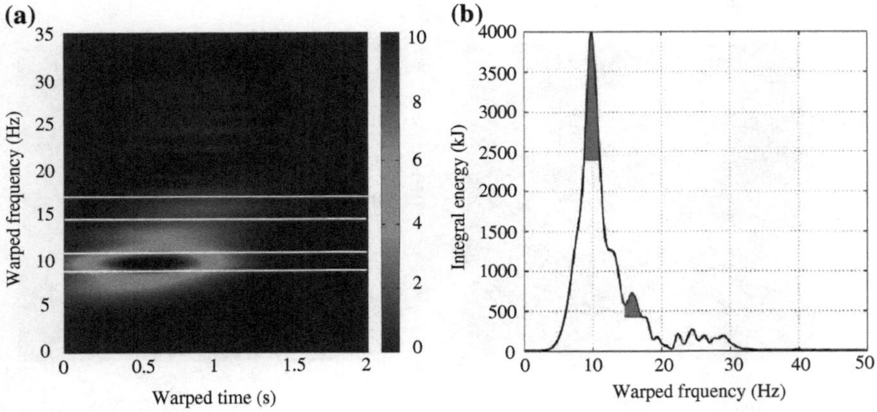

Fig. 3 The spectrogram of the warped signal $W_h^t p(t_w)$ (**a**) and its integral energy (**b**). We find all the sufficiently high energy maxima and select stripes where the integral energy exceeds 65 % of the value at the maximum. These stripes are highlighted in *gray* on the integral energy plot (**b**), and their boundaries are indicated by the *white lines* on the spectrogram

The spectrogram $SP(W_h^t p)(t_w,f_w)$ of the warped signal is shown in Fig. 3a. We computed the integral energy for every frequency by the formula

$$I(f_w) = \int_0^{t_{\max}} SP(W_h^t p)(t_w,f_w). \tag{3}$$

The plot of $I(f_w)$ as a function of f_w is shown in Fig. 3b. Every mode corresponds to a maximum of $I(f_w)$. A stripe corresponding to the values of the integral energy exceeding 65 % of each maximum is used to separate the parts of the spectrogram corresponding to different modes. These filtered modes are then reverse transformed to the "original" time domain and after some reassignment processing, we can plot the extracted dispersion curves against the spectrogram of the original signal (Fig. 4). This spectrogram is plotted in linear (Fig. 4a) and logarithmic (Fig. 4b) intensity scale in order to show the contribution of the weaker spectral components of the original signal.

The dispersion curves seem to be reasonably located on the spectrogram $SP(p)(t, f)$. Note that the energy on the spectrogram of the signal $p(t)$ is concentrated around the frequency of 73.5 Hz. In the vicinity of this frequency, there are two waterborne modes and the most energy is propagated mainly by the first one. Now we are ready to perform the inversion via minimizing the cost function Eq. (2).

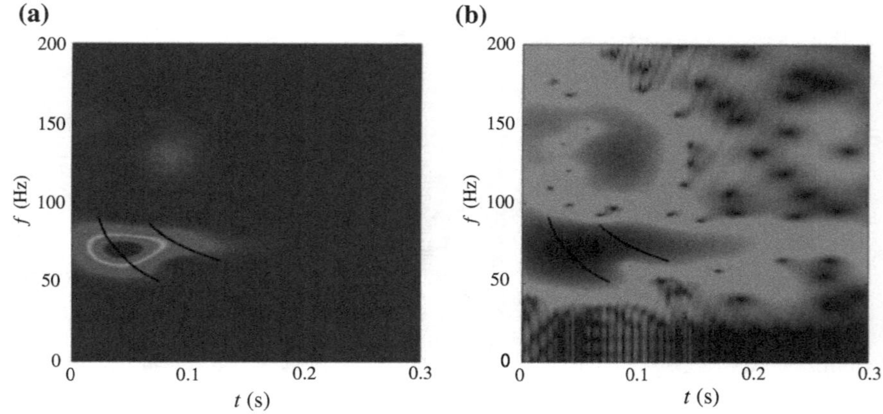

Fig. 4 The spectrogram $SP(p)(t, f)$ of the original signal $p(t)$ plotted for the linear intensity scale **a** and logarithmic scale **b**. The recovered dispersion *curves* are plotted against the spectrogram. Note the poor modal structure of the signal especially in comparison to the signals from Refs. [1, 2]

4 Results and Discussion

The inversion search space is described in Table 1. The maximal and minimal values of the sound speed and density were taken from the existing data of the geological research in this area [5] (only very rough estimates are given there). The resulting values of the inverted parameters are shown in the last column. The resulting value of the bottom depth may be considered as indirect confirmation of the estimation validity.

This allows us to conclude that the warping-based single-hydrophone inversion scheme can be successfully applied to the signals containing very small amount of the dispersion data. At present, the method of warping-based inversion is being intensively developed and in our view many things still have to be improved.

Table 1 The search space of the geoacoustic inversion scheme and the resulting values of the bottom parameters c_b, ρ_b, H and the time shift $d\tau$

Parameter	Min. value	Max. value	Step	Inversion result
c_b (m/s)	1550	1850	5	1720
ρ_b (g/cm^3)	1.5	2	0.05	1.7
H (m)	20	40	0.5	27
$d\tau$	−0.03	0.03	0.001	0.013

Acknowledgments We are very much obliged to Dr. Julien Bonnel and Prof. Ross Chapman for their kind attention to our numerous questions concerning the warping-based inversion method, for sharing their experience and parts of their code. We are also very grateful to Dr. Alexander Rutenko, Dr. Alexander Solovyev and Denis Manul'chev for providing us with the experimental data and patiently answering our questions related to the conduct of their experiments. The reported study was partially supported by RFBR (research project No. 12-05-33022-mol_a_ved).

References

1. Bonnel, J., Chapman, N.R.: Detection capacities of distributed and centralized systems: a comparative study. J. Acoust. Soc. Am. **130**(2), EL101-7 (2011)
2. Bonnel, J., Gervaise, C., Nicolas, B., Mars, J.I.: Single-receiver geoacoustic inversion using modal reversal. J. Acoust. Soc. Am. **131**(1), 119–128 (2012)
3. Jensen, F.B., Porter, M.B., Kuperman, W.A.H.: Schmidt Computational Ocean Acoustics. Springer, New York (2011)
4. Tolstoy, I., Clay, C.: Ocean Acoustics: Theory and Experiment in Underwater Sound. Acoustical Society of America, New York (1987)
5. Korotchenko, A.N., Samchenko, A.N., Yaroshchuk, I.O.: Vestnik DVO RAN **6**(1), 54–59 (2011) (in Russian)

Measurements of Ultrasound Attenuation of Suspended Particles with Various Size Distributions

Changil Lee, Jee Woong Choi and Jungyul Na

Abstract Acoustic attenuation in suspended sediments can be considered as a sum of viscous and scattering absorptions, which are strongly affected by a distribution in sediment particle size. Generally, attenuation models to predict attenuation in suspended sediment use a mean particle size as an input parameter. However, because the suspended particles in the ocean exist not just as a form of a single particle size, but with a particle-size distribution, it may cause an error between the model prediction and measurements. In this paper, sound attenuation measurements for suspended sediments were conducted using 5 and 7.5 MHz acoustic signals, for which several suspended sediment samples with different size distributions and concentrations were utilized. Measurement results are presented here and their comparisons to the model predictions are also discussed.

Keywords Suspended sediment · Ultrasound attenuation · Particle-size distribution

1 Introduction

The effects on sound attenuation in the ocean can be divided by the absorption loss of the water medium itself and the loss due to sediment particles suspended in the water medium. The attenuation in the suspended sediment is again considered as a sum of viscous absorption loss and scattering loss [1, 2]. Although suspended sediment is characterized by a particle-size distribution, the sound attenuation is generally pre-

C. Lee (✉) · J.W. Choi · J. Na
Department of Marine Sciences and Convergent Technology, Hanyang University, Gyeonggi-do 426-791, Korea
e-mail: aejan17@gmail.com

J.W. Choi
e-mail: choijw@hanyang.ac.kr

J. Na
e-mail: najy0252@hanyang.ac.kr

L. Zhou et al. (eds.), *Underwater Acoustics and Ocean Dynamics*
DOI 10.1007/978-981-10-2422-1_6

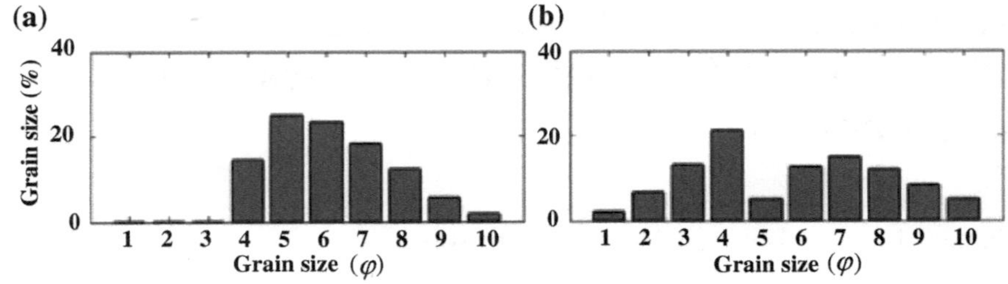

Fig. 1 Particle-size distributions of **a** kaolinite powder and **b** pink-kaolinite powder

dicted using a single particle size, representing its distribution, such as a mass-median diameter, D50, which may give rise to error. In this paper, the acoustic attenuation at high-concentration suspensions with different size distributions and concentrations are measured and the results are compared to the model predictions.

2 Laboratory Measurements

Attenuation measurements using ultrasound signals were performed on the artificial suspended sediment samples made by kaolinite and pink-kaolinite powders. A total of four ultrasound transducers were used and each two transducers were aligned face to face in a PVC water tank, each comprising a transmitter and receiver pair. The concentration of suspended sediments was controlled to range from 0 to 19 g/L, and CW acoustic signals with center frequencies of 5 and 7.5 MHz were used as source signals. Figure 1 shows the particle-size distributions of kaolinite and pink-kaolinite powders. The values of D50 for the kaolinite and pink-kaolinite powders were estimated to be 24 and 34 μm, respectively, which were used as input parameters for the attenuation model.

3 Results

Figure 2 shows comparisons between the attenuation coefficients obtained by model predictions and measurements for two kinds of suspended sediment samples. For the case of 5 MHz, overall the measurements reasonably agree well with the model predictions for both suspended sediment samples. However, for the case of 7.5 MHz, the difference between the model prediction and measurements increases as the concentration increases, especially for the pink-kaolinite case. Note that the distribution of pink-kaolinite powder is characterized by a bimodal distribution although it was plotted as a function of the phi scale of the particle size, as shown in Fig. 1b. The results imply that D50 may not be useful for characterizing the

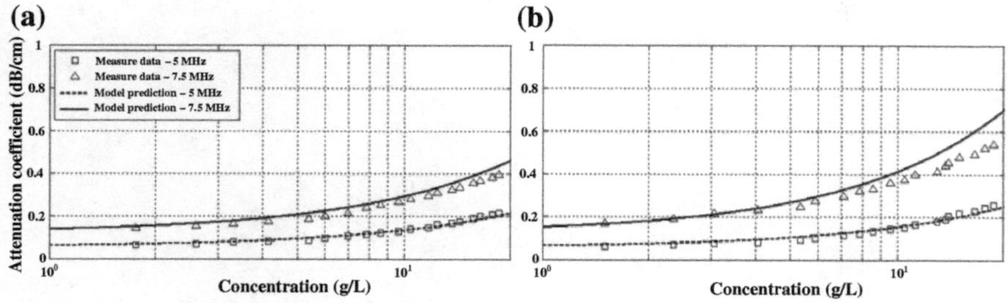

Fig. 2 Comparisons between attenuation measurements and model predictions using D50 as a model input for **a** kaolinite sample and **b** pink-kaolinite sample

distribution in particle size. Further study, including an investigation of the effect of particle-size distribution on the attenuation, is needed.

Acknowledgments This work was supported by the National Research Foundation of Korea (No. 2010-0022815).

References

1. Richard, S.D., Heathershaw, A.D., Thorne, P.D.: The effect of suspended particulate matter on sound attenuation in seawater. J. Acoust. Soc. Am. **100**(3), 1447–1450 (1996)
2. Ha, H.K., Maa, J.P.Y., Park, K., Kin, Y.H.: Estimation of high-resolution sediment concentration profiles in bottom boundary layer using pulse-coherent acoustic Doppler current profiles. Mar. Geol. **279**, 199–209 (2011)

A Low Complexity Multichannel Adaptive Turbo Equalizer for a Large Delay Spread Sparse Underwater Acoustic Channel

Yanbo Wu and Min Zhu

Abstract A scheme for a multichannel adaptive turbo equalizer for a large delay spread sparse underwater acoustic channel is presented, with the added complexity of a single DSP chip implementation. A channel equalizer without an input of a decoder's soft information is structured as a multichannel feed-forward (FF) filter with a phase lock loop (PLL), and during the iterative processing with a decoder, an inter-symbol interference canceller (IC) filter is added in the channel equalizer structure to further utilize the soft information of the decoder. An FF filter and an IC filter are sparselized based on a magnitude truncation of full order equalizer filters. Multichannel waveforms in the training period are used to calculate filter taps with a non-sparse least mean square (LMS) adaptive algorithm, with an order larger than 1000 (for example: 4 receiving channels, 2 samples per symbol each channel, FF filter duration is 160 symbols, then the multichannel FF filter order is 1280, and the real-time DSP chip implementation is possible with an LMS algorithm). After sparse tap position selection, the order of the equalizer filter is reduced to less than 200. Multichannel receiving waveforms of both training and data symbols are processed with a fast self-optimized LMS (FOLMS) adaptive sparse equalizer with PLL. A turbo code, bit-interleaving coded modulation (BICM), and QPSK/8PSK with gray mapping are adopted as they perform well in a fading channel. The information exchanges between the equalizer and decoder are realized by symbol/bit converters. The scheme is robust to combat delay spread with a duration larger than a 100 symbols and can easily be implemented on a single DSP chip. An extrinsic information transfer (EXIT) chart is used to analyze the performances of a sparse equalizer, a channel decoder and their iteration. Shallow lake and ocean experiments verified the performance of the scheme, with a PSK symbol rate of 4k Baud, and half rate turbo code. Communication distance was from 1 to 5 km, and

---

Y. Wu (✉) · M. Zhu
Ocean Acoustic Technology Laboratory, Institute of Acoustics,
Chinese Academy of Sciences, Beijing 100190, China
e-mail: wuyanbo@mail.ioa.ac.cn

Y. Wu · M. Zhu
State Key Laboratory of Acoustics, Institute of Acoustics,
Chinese Academy of Sciences, Beijing 100190, China

L. Zhou et al. (eds.), *Underwater Acoustics and Ocean Dynamics*
DOI 10.1007/978-981-10-2422-1_7

there was no error frame with the iteration of the turbo equalization, even when the delay spread was larger than 40 symbols and with no dominant channel path.

Keywords Underwater acoustic communication · Sparse channel · Large delay spread · Turbo equalizer

1 Introduction

The characteristics of an underwater acoustic communication channel are sparse, time-varying, and multipath, especially in long-distance shallow water [1, 2]. A multipath is caused by the reflection of the surface or bottom, and a curving sound ray. The variability of the amplitude and phase of the channel impulse response is mainly caused by movement of communication devices. When communication devices are lowered from floating ships, rather than being moored from the water bottom, the channel variability is especially obvious. The difficulty of shallow water communication is to confront the varying long multipath, and associated reverberation. A multichannel adaptive equalizer with phase tracking is a classic method [1], which can work well in a sufficient signal-to-noise ratio (SNR) environment. A turbo equalizer [3], using the turbo iteration principle, decreases the SNR requirement. An adaptive turbo equalizer was studied in Ref. [3], to track the variability of an underwater channel [4]. Both the adaptive equalizer and adaptive turbo equalizer face the problem of a large number of equalizer taps, which decreases the tracking speed, and increases the misadjustment noise and the computation complexity. A sparse underwater channel estimation was studied in Ref. [5]. However, the optimum tap selection of the equalizer based on a sparse channel estimation is rather complicated, especially with multichannel and soft decision feedback [2]. This paper introduces a low complexity method to realize an adaptive sparse turbo equalizer. The sparse tap selection is simple but efficient. A turbo equalizer with 4 receiving channels is finally realized in real-time on a single DSP chip, which can adaptively deal with 1–5 km distances with different multipath channel structures.

2 Proposed Sparse Equalizer

To utilize the benefit of sparse characteristics, several considerations should be taken into account. First, sparse position is constant during one frame. The frame duration used in the proposed scheme is less than 1 s, during which there is no new path emerging. Second, each path is a cluster, not a single ray. For the channel impulse response, the sparse position is not isolated but continuous with several symbols. Third, sparse position needs to be jointly optimized among different receiving array elements. With these considerations, the proposed method of sparse position initialization contains the following steps as shown in Fig. 1. Step 1:

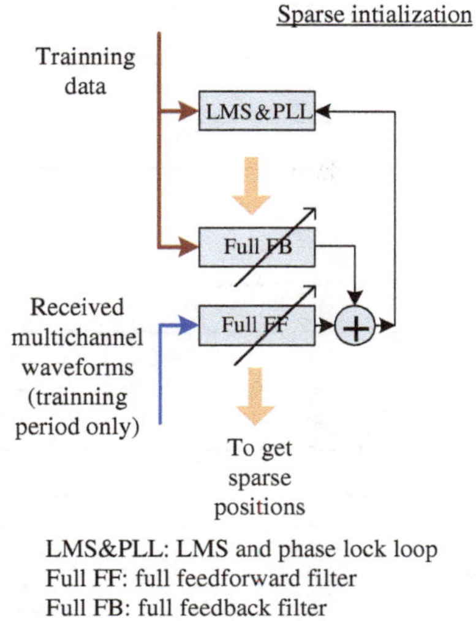

Fig. 1 Proposed method of sparse position initialization

ensure equalizing adaptively during training period with full taps. The tap length is constant according to the maximum delay spread duration, and the least mean square (LMS) algorithm is used as it has advantages of robustness and low complexity. Step 2: find positions of equalizer taps by magnitude truncation. A constant ratio to maximum amplitude of taps is used. Step 3: expand selected positions (to ± 3 nearby positions). Figure 2 shows 4 channel equalizer taps and sparse positions, with a distance of 1470 m. The full order number is 1288, which is reduced to 148 by sparse selection.

3 Adaptive Turbo Equalizer with Sparse Taps

An adaptive multichannel linear turbo equalizer is used with the basic idea shown in Ref. [3]. With the first equalization, the linear equalizer without feedback is adopted to avoid error propagation. During the iteration of the turbo equalization, soft output from the decoder is fed into the feedback equalizer filter with both post cursor and precursor. The proposed method here has the following differences from others, as shown in Fig. 3. The equalizer is a sparse filter structure. The input data of the equalizer is not circular after sparse selection, so the reduced complexity RLS cannot be used. Meanwhile, the order of the input data is also a variable which depends on the channel spread length; the step factor must be adaptively selected. A fast optimized LMS is preferred, and a normalized LMS may be suitable for real-time implementation. For the channel code, a turbo code is used, as less iterations (3–5 are sufficient) of turbo equalization are needed compare to an RSC code

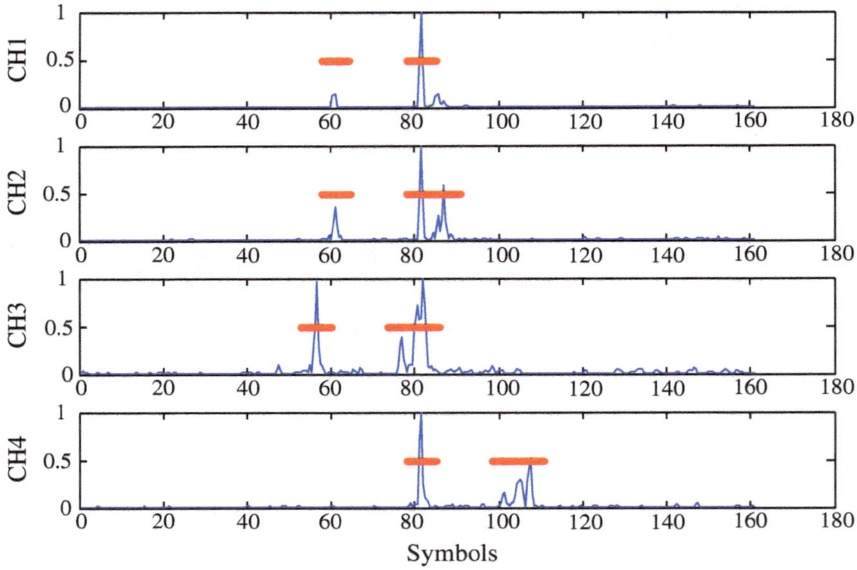

Fig. 2 Four channel equalizer taps and sparse position

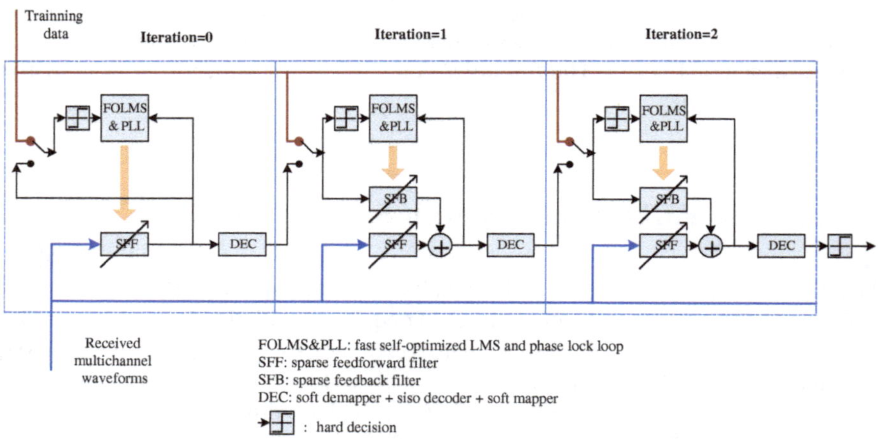

Fig. 3 Structure of proposed adaptive turbo equalizer with sparse taps

(10–20). Another reason is that the turbo code performs better with an AWGN channel than with an RSC code.

4 Shallow Sea Experiment

A shallow sea communication experiment was carried out in South China Sea, in August and September 2013. The sound speed profile is shown in Fig. 4. The water depth was 60 m, and 1 transducer and 4 receiving hydrophones were lowered from two ships with a depth of about 30 m. A QPSK was used with a symbol rate of 4k Baud, and the carrier frequency was 8 kHz. A 1/2 turbo code was used, with a frame length of 1936 bits. A channel impulse response and extrinsic information transfer (EXIT) chart [6] of different distances is shown in Figs. 5, 6 and 7. In the EXIT chart, the mutual information curves of the equalizers (without/with feedback) and the channel decoder are shown in red or blue lines. From the figures, we can see that the channel with a moderate distance is favorable, and the SNR requirement is low; however, the advantage of iteration is not so obvious.

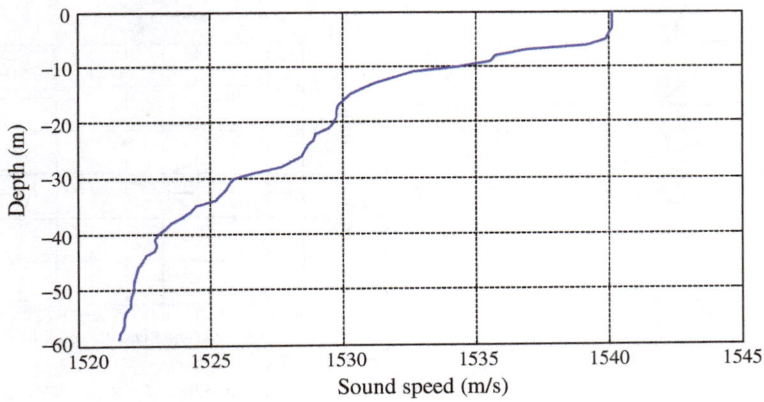

Fig. 4 Sound speed profiles in South China Sea (Sept. 3rd, 2013)

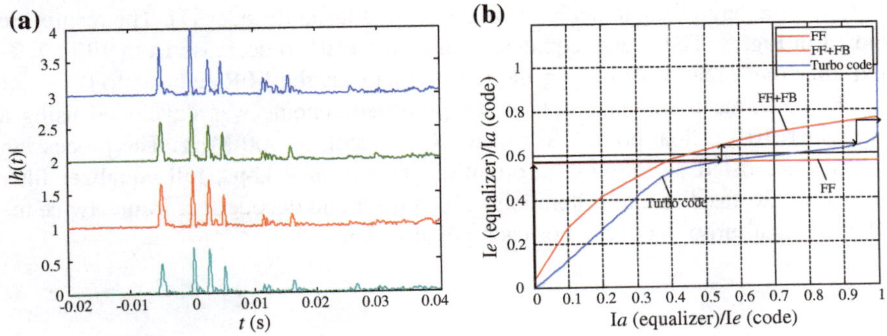

Fig. 5 Channel impulse response and EXIT chart (1) (Distance = 1159 m, SNR = 1 dB/channel)

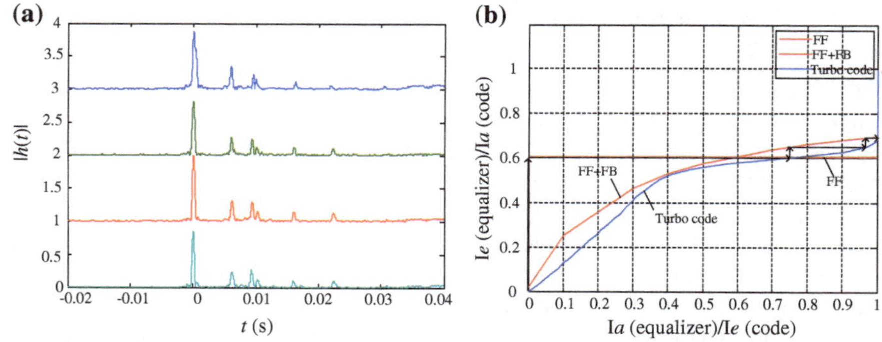

Fig. 6 Channel impulse response and EXIT chart (2) (Distance = 2421 m, SNR = −1.5 dB/channel)

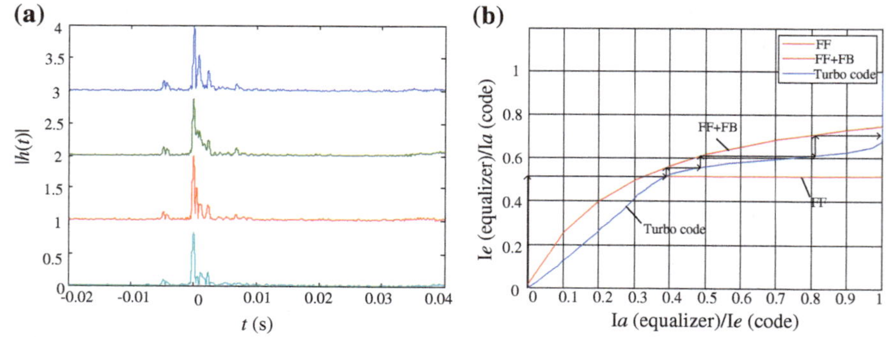

Fig. 7 Channel impulse response and EXIT chart (3) (Distance = 4258 m, SNR = −0.5 dB/channel)

1624 frames were collected, with varying distances from 1 to 5 km. The proposed scheme (equalizer spans 160 symbols) was compared to a traditional adaptive DFE scheme (equalizer spans 32 symbols) with a turbo decoder [7]. The results are shown in Fig. 8. The sparse equalizer causes the FER to decrease from 19 to 7 %, and after five iterations of using the turbo equalizer, the FER reduces to 0.

The real-time implementation of the proposed scheme was developed using a TigerSharc101, a float-point DSP, with a CPU clock of 300 MHz. The processing abilities are listed as follows: information bit rate of 4 kbps, full equalizer filter order 640 (20 ms 4 ch), four iterations of equalizer and decoder per frame, twice the iteration of a turbo decoder after each equalization.

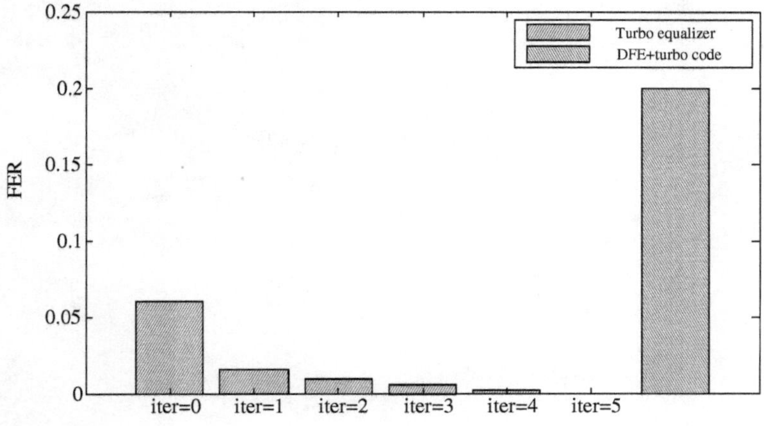

Fig. 8 Frame error rate comparisons

Acknowledgments This work was supported by the Chinese National "863" Program (grant Nos. 2009AA093301 and 2009AA093601).

References

1. Stojanovic, M., Catipovic, J.A., Proakis, J.G.: Phase coherent digital communications for underwater acoustic channels. IEEE J. Ocean. Eng. **19**(1), 100–111 (1994)
2. Kocic, M., Brady, D., Stojanovic, M.: Sparse equalization for real-time digital underwater acoustic communications. In: OCEANS'95. MTS/IEEE. Challenges of Our Changing Global Environment. Conference Proceedings, vol. 3, pp. 1417–1422. IEEE (1995)
3. Laot, C., Glavieux, A., Labat, J.: Turbo equalization: adaptive equalization and channel decoding jointly optimized. IEEE J. Sel. Areas Commun. **19**(9), 1744–1752 (2001)
4. Choi, J.W., Riedl, T.J., Kim, K., Singer, A.C., Preisig, J.C.: Adaptive linear turbo equalization over doubly selective channels. Oceanic Eng. **36**(4), 473–489 (2011)
5. Cotter, S.F., Rao, B.D.: Sparse channel estimation via matching pursuit with application to equalization. IEEE Trans. Commun. **50**(3), 374–377 (2002)
6. Lee, S.J., Singer, A.C., Shanbhag, N.R.: Analysis of linear turbo equalizer via EXIT chart. In: Global Telecommunications Conference, GLOBECOM'03, vol. 4, pp. 2237–2242. IEEE (2003)
7. Zhu, W.Q., Zhu, M., Wu, Y.B., et al.: Signal processing in underwater acoustic communication system for manned deep submersible "Jiaolong". Chin. J. Acoust. **32**(1), 3238 (2013)

A Turbo Equalization Based on a Sparse Doubly Spread Acoustic Channels Estimation

Zibin Yu, Hangfang Zhao, Wen Xu and Xianyi Gong

Abstract This paper proposes a novel Turbo equalizer structure that can operate reliably and efficiently over time delay and Doppler doubly spread acoustic channels. It is based on the compressive sensing channel estimation method of the sparse doubly spread channels.

Keywords Turbo equalization · Underwater acoustic communications (UAC) · Doubly spread channels

1 Introduction

The difficulties of underwater acoustic communications (UAC) are mainly due to the complexity of underwater acoustic channels. Because of multiple reflections, refractions, and scattering acoustic waves are prone to propagating along multiple paths, which leads to a large but sparse delay spread in channel impulse response (CIR). On the other hand, CIR is time varying because of the transmitter/receiver motions, reflection caused by the fluctuating sea surface and scattering of internal waves, which leads to a Doppler spread [1]. Hence, such channels are characterized by their large but sparse time delay spread and Doppler spread, which are referred to as time delay and Doppler doubly spread (TDDS) channels. Because of TDDS channels, traditional systems that are based on time invariant CIR assumptions often perform poorly for UAC.

Z. Yu (✉) · H. Zhao · W. Xu · X. Gong
College of Information Science and Electronic Engineering,
Zhejiang University, Hangzhou 310027, China
e-mail: hfzhao@zju.edu.cn

W. Xu
e-mail: wxu@zju.edu.cn

X. Gong
Hangzhou Applied Acoustic Research Institute, Hangzhou 310027, China

L. Zhou et al. (eds.), *Underwater Acoustics and Ocean Dynamics*
DOI 10.1007/978-981-10-2422-1_8

In order to achieve reliable communications under channel uncertainties, coding gain must be utilized to improve communication systems performance in low SNR regions. Turbo equalization [2], which is a natural extension of turbo codes, views a channel with inter symbol interference (ISI) as an inner convolutional encoder and utilizes serial concatenated convolutional codes with interleaver to greatly decrease the occurrences of low weight code words. Hence it can achieve excellent performance in low SNR regions, which is very close to the theoretical bound predicted by Shannon. In the receiver end, turbo equalization utilizes probability decoding and iterative soft information exchange between the constitute channel equalizer and the channel decoder so as to obtain the coding gain. Hence, both the decoder and the equalizer should be soft input and soft output (SISO). A common choice for a SISO equalizer is the maximum a posteriori probability (MAP) equalizer. But for UAC, there are two difficulties for the implementation of turbo equalization. The first difficulty is, due to the large delay spread, the MAP equalizer with exponential algorithm complexity is impractical for implementation. In order to reduce the algorithm complexity, an LMMSE equalizer is modified into the SISO form so as to replace the MAP equalizer. The second difficulty is the time varying property of CIR. Traditional turbo equalizers with an LMMSE channel equalizer assume a time invariant CIR and hence frequently a channel estimation update is required to hold the time invariant CIR assumption [3], which is certainly a waste of computational resources and available bandwidth resources.

To overcome those two difficulties, this paper replaces the LMMSE channel equalizer based on CIR estimation with an LMMSE channel equalizer based on sparse TDDS channel estimation. That is based on the observation that the TDDS channel estimation can be regarded as a first-order approximation to the rapidly time varying channel and is much stabler than a CIR [1]. For this reason, a Turbo equalizer is modified to provide a fit for the embedding of a TDDS channel estimation. With the much more stable TDDS channel estimation in hand, the system proposed in this paper can provide a quite satisfactory performance for a TDDS channel without requiring frequent channel estimation updates.

2 Theory

In order to combine the turbo equalizer structure with the doubly spread channel, we note that the received data can be written as Eq. (1)

$$y_n = H_n x_n + w_n \tag{1}$$

where $y_n = [y_{n-W_2}, y_{n-W_2+1}, \ldots, y_{n+W_1}]^{\mathrm{T}}$, $x_n = [x_{n-W_2+M+1}, x_{n-W_2+M+2}, \ldots, x_{n+W_1}]^{\mathrm{T}}$, M is the maximum number of the time delay tap.

$$H_n = \sum_l \Phi_l(n) U_l(n) \tag{2}$$

where

$$\Phi_l(n) = \begin{bmatrix} e^{j2\pi v_l (n-W_2)} & 0 & \cdots & 0 \\ 0 & \ddots & \ddots & \vdots \\ \vdots & \ddots & \ddots & 0 \\ 0 & \cdots & 0 & e^{j2\pi v_l (n+W_1)} \end{bmatrix} \tag{3}$$

$$U_l(n) = \begin{bmatrix} U[M-1,l] & U[M-2,l] & \cdots & U[0,l] & 0 & \cdots & 0 \\ 0 & U[M-1,l] & \cdots & U[1,l] & U[0,l] & \cdots & 0 \\ \vdots & \vdots & \vdots & \vdots & \vdots & \vdots & \vdots \\ 0 & \cdots & 0 & U[M-1,l] & \cdots & U[1,l] & U[0,l] \end{bmatrix} \tag{4}$$

By incorporating Eqs. (1)–(4) into the wide sense LMMSE equalizer of the transmitted symbols, one can easily derive the LMMSE equalizer under the doubly spread channel as follows:

$$\hat{x}_n = \mu_n + u_n (y_n - H_n \mu_n) + v_n (y_n - H_n \mu_n)^* \tag{5}$$

where

$$u_n = \left(v_n h_n^{\mathrm{H}} - \psi_n h_n^{\mathrm{T}} (\Sigma_n^{-1})^* (H_n \psi_n H_n^{\mathrm{T}})^*\right) R_n \tag{6}$$

$$v_n = \left(\psi_n h_n^{\mathrm{T}} - \psi_n h_n^{\mathrm{H}} \Sigma_n^{-1} H_n \psi_n H_n^{\mathrm{T}}\right) R_n^* \tag{7}$$

$$R_n = \left(\Sigma_n - H_n \psi_n H_n^{\mathrm{T}} (\Sigma_n^{-1})^* (H_n \psi_n H_n^{\mathrm{T}})^*\right)^{-1} \tag{8}$$

$$\mu_n = \sum_{x \in \mathcal{S}} x \cdot s'(x_n = x) \tag{9}$$

$$v_n = \sum_{x \in \mathcal{S}} |x - \mu_n|^2 \cdot s'(x_n = x) \tag{10}$$

$$\psi_n = \mathrm{Cov}\{x_n, x_n^*\} = \sum_{x \in \mathcal{S}} (x - \mu_n)^2 \cdot s'(x_n = x) \tag{11}$$

$$\Sigma_n = \sigma^2 I_W + H_n V_n H_n^{\mathrm{T}} \tag{12}$$

$$V_n = \mathrm{Diag}\{v_{n-M_h-W_2}, v_{n-M_h-W_2+1}, \ldots, v_{n+W_1}\} \tag{13}$$

$$\Psi_n = \mathrm{Diag}\{\psi_{n-M_h-W_2+1}, \psi_{n-M_h-W_2+2}, \ldots, \psi_{n+W_1}\} \quad (14)$$

The decoder structure is the same as usual [2]. In every turbo iteration, it can provide our proposed equalizer with soft information $s'(x_n = x)$. By assuming that the error between the equalizer output and the true solution obeys the Gaussian distribution, we can obtain the soft information from our proposed LMMSE equalizer to the decoder. Hence, the algorithm of the turbo equalizer over doubly spread channel is obtained.

3 Simulation Results

Simulation results show that for a UAC with a large delay and Doppler spread, the traditional turbo equalizer based on CIR estimation fails to achieve reliable communication while our modified turbo equalizer based on a TDDS channel estimation performs quite well. Figure 1a depicts the time evolution of a simulated

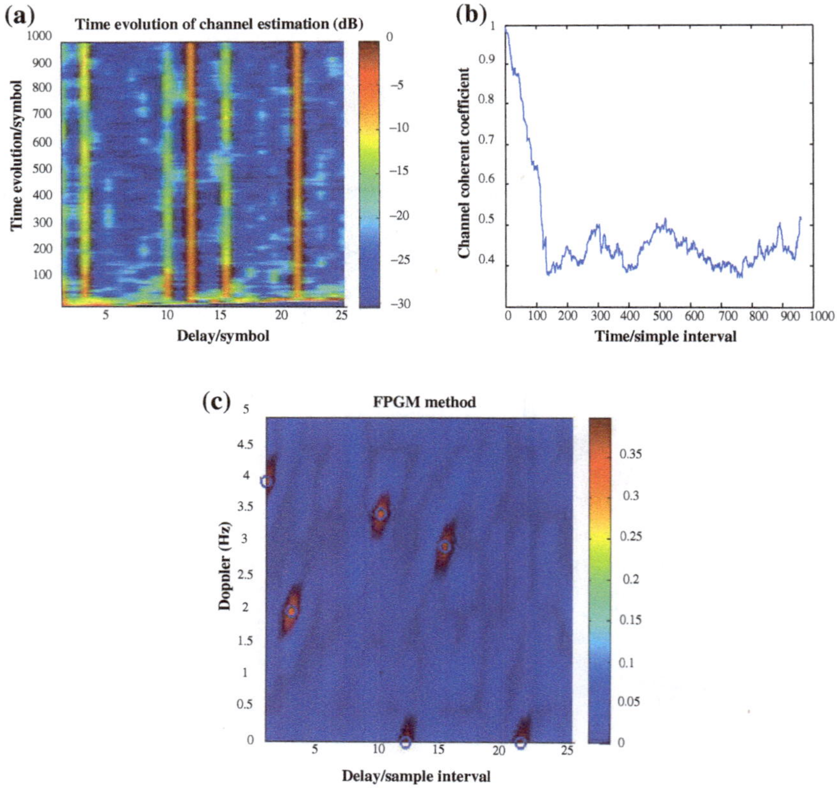

Fig. 1 Time evolution of CIR for TDDS channel **a**, time evolution of channel coherence coefficient of TDDS channel **b** and TDDS channel estimation of FPGM method **c**

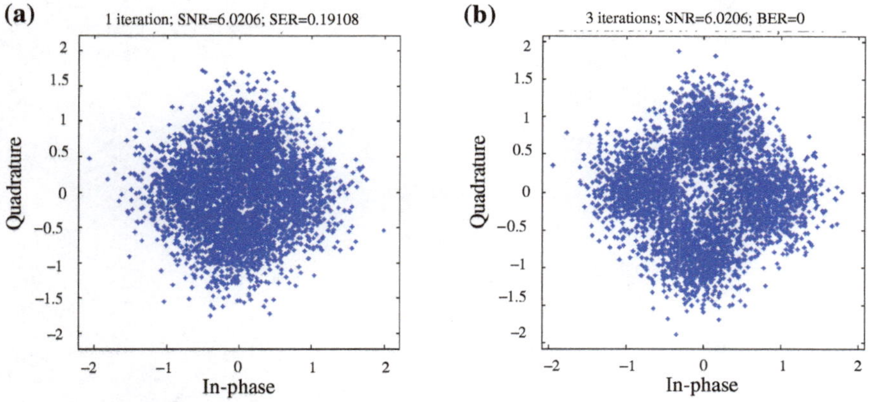

Fig. 2 Turbo equalization output with 1 iteration, symbol error rate (SER) = 0.19108 (**a**) and three iterations, bit error rate (BER) = 0 (**b**)

classical fading channel with five main paths for different Doppler frequency shifts. Figure 1b shows that the coherence time (time for channel coherence coefficient to be above 0.8) is less than 100 sample intervals, which means that the traditional turbo equalization method has to update the CIR estimation every 100 sample intervals (quite impractical). Figure 1c is the TDDS channel estimation result with the FPGM method in Ref. [1]. By utilizing the sparsity of the TDDS channel, only a small number of training sequences are required to obtain a reliable TDDS channel estimation. In our simulation, 4082 QPSK communication signals are transmitted through the fading channel of Fig. 1 with a relatively low SNR (6.0206 dB in our example) so as to be distinguished from the results presented in Ref. [1]. Only one TDDS channel estimate is obtained from the initial time with 200 training sequences. Our proposed turbo equalizer's outputs are depicted in Fig. 2. Figure 2a is the result with 1 iterative extrinsic information exchange between channel equalizer and channel decoder while the result with three iterations is shown in Fig. 2b. From Fig. 2, we can obtain the conclusion that in a TDDS channel, our proposed turbo equalizer can achieve a 0 bit error rate (BER) communication under low SNR with only three iterations and requires no channel update for quite a long time (4082 sample intervals).

References

1. Zeng, W.J., Xu, W.: Fast estimation of sparse doubly spread acoustic channels. J. Acoust. Soc. Am. **131**(1), 303–317 (2012)
2. Tuchler, M., et al.: Turbo equalization: principles and new results. IEEE Trans. Commun. **50**(5), 754–767 (2002)
3. Choi, J.W., et al.: Adaptive linear turbo equalization over doubly selective channels. IEEE J. Oceanic Eng. **36**(4), 473–489 (2011)

Design and Testing of Underwater Acoustic Communications for an AUV

Hongtao Zhang, Zhijie Yue, Zhe Xie and Zhongkang Wang

Abstract This paper presents an autonomous underwater vehicle (AUV) underwater acoustic communication system. There are two communication modes integrated to meet the different requirements, that is, (1) spread spectrum underwater acoustic communication, with a variable transmission rate from 70 to 400 bps, to transmit instructions, (2) orthogonal frequency division multiplexing (OFDM) communication, with a variable transmission rate from 600 bps to 2.4 kbps, to transmit data information. The communication mode can be distinguished according to the sweep frequency direction of the Linear Frequency Modulation (LFM) chirp signal in the receiver. A four-element hydrophone array is lowered from the mother ship to a certain depth, and passive time reversal technology is adopted, the array element spacing is 4 times the wavelength of a signal (6–10 kHz bandwidth) at the center frequency. The symbol timing offset and Doppler shift caused by moving between the AUV and the mother ship were simultaneously estimated and corrected. The AUV communication lake test and sea trial were performed in August and September 2013, and the results of the experiments are reported in this paper.

Keywords Underwater acoustic communications · AUV · Modem

1 Introduction

The effective exchange of information is desirable for an autonomous underwater vehicle (AUV) operation, including the downlink control instructions and uplink AUV running status. When operating at the surface, various air communication technologies, mostly based on radio frequencies, can be employed to easily transfer data between the topside control and the monitoring system and the submerged AUV. These approaches offer good communication range and high bandwidth. Once

H. Zhang (✉) · Z. Yue · Z. Xie · Z. Wang
Hangzhou Applied Acoustics Research Institute, Hangzhou 310027, China
e-mail: zhtcsic@126.com

L. Zhou et al. (eds.), *Underwater Acoustics and Ocean Dynamics*
DOI 10.1007/978-981-10-2422-1_9

submerged, as radio frequencies do not propagate well in seawater, information transfer between the AUV and the operator becomes both critical and more difficult.

This paper describes an underwater acoustic communications system to exchange data and information between a submerged AUV and the operator. This work was supported by the National "863" Program. The goal of this study was to develop an acoustic underwater communications system that would enable reliable underwater acoustic communication.

Depending on the communications system scenario, there are different requirements of transmission speed and reliability in an acoustic link. It may be slow but reliable transmission of the AUV's control signals, faster and still reliable measurement data transmission from an underwater monitoring system or video transmission as fast as possible from underwater cameras. Two communication modes, including the spread spectrum communication and orthogonal frequency division multiplexing (OFDM) communication are integrated to meet the different requirements for communicating.

2 Spread Spectrum Acoustic Communication

The spread spectrum method can provide a reliable, low bit rate and long-range underwater acoustic link with low probability of intercept. It demonstrates a good performance of suppression of multi-path interference in underwater acoustic channels [1–3]. In this project, a direct sequence spread spectrum communication scheme is designed to transmit control instructions.

2.1 Signals Design

Figure 1 shows the simple framework for this scheme. Each symbol contains two sequences: one is a reference pseudo-noise (PN) sequence used to precisely estimate the delay and Doppler, and the other is information PN sequence used to carry information. These two sequences are modulated into an in-phase and a

Fig. 1 DSSS transmit signals

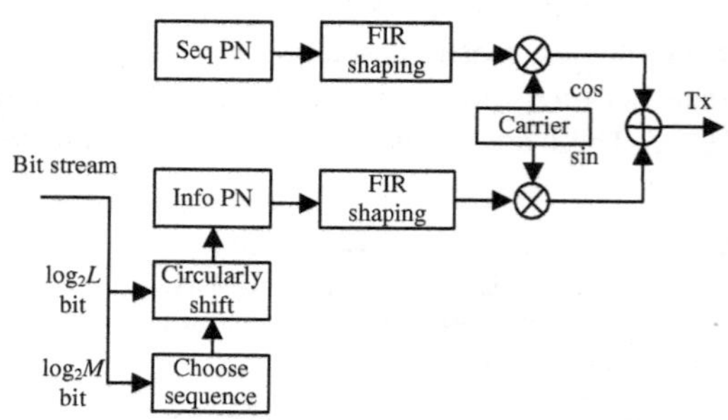

Table 1 DSSS bit rate

L	128	256	512	1024
Bit rate (bps)	437	250	140	78

quadrature-phase branch, respectively. Following the PN spreading, these two quadrature signal components pass through identical pulse shaping filters, which are designed to have a raised-cosine spectrum. The length of the information sequence is L, which can be chosen from a sequence set containing the M sequence. The information bit stream is divided into small sections, each section consisting of $\log_2 M$ and $\log_2 L$ bits; $\log_2 M$ bits decide which sequence is chosen from the group, and $\log_2 L$ bits decide the position of the circular shift, so one symbol can carry $\log_2 M$ and $\log_2 L$ bits. We choose the orthogonal gold sequence as the PN sequence. The sequence length L varies from 128 to 1024, so one symbol can carry 14–20 bits. Table 1 shows the bit rate of different sequence lengths in the case of a 4 kHz chip rate.

2.2 *Communication Modes*

A Linear Frequency Modulation (LFM) signal is added before the information signals. If the sweep frequency direction of the LFM signal is from low to high, the information signal is DSSS; conversely, if the sweep frequency direction is from high to low, the information signal is OFDM. The communication mode can be distinguished according to the sweep frequency direction of the LFM signal by matching the filter in the receiver.

2.3 *Doppler Estimation and Compensation*

The receiver utilizes the LFM to estimate the first coarse symbol timing. The fine symbol timing and Doppler spread can be estimated by an ambiguity function. Because the AUV maximum moving speed is known, the maximum time- and frequency spread value in each symbol period can be predicted. The search step in the time and frequency domain is decided by the resolution in the time and frequency domain, the Doppler range and DSP process ability. This search can be realized by 2D discrete correlation of different delay and resample replicas of the received signal with the known reference PN sequence. Upon synchronization of the time delay and resample rate, the correlator yields the largest correlation peak. In order to increase the speed of the calculation, a fast Walsh transform (FWT) is used in the correlation operation.

3 OFDM Acoustic Communication

OFDM has certain advantages including high spectral efficiency and reliable performance against inter-symbol interference caused by multi-path propagation, so it attracts more attention in high-bit-rate UWA communication [4–6]. In this project, OFDM is designed to transmit information data.

3.1 Signals Design

We use a zero-padded (ZP) OFDM with a guard interval of 1/4 symbol duration. Let T denote the symbol duration and T_g the guard interval. The total OFDM block duration is $T_0 = T + T_g$ and the subcarrier spacing is $1/T$. The number of the subcarrier is K, so that the bandwidth is $B = K/T$. The K subcarriers consist of the K_g guard subcarriers, K_p pilot subcarriers, and K_a information data subcarriers. Guard subcarriers are allocated at the upper and lower band edges. Pilot subcarriers are used for Doppler estimation and channel estimation. Figure 2 shows the arrangement of subcarriers. For instance, the total of subcarriers K is 512, and the upper and lower guard subcarriers are 32, respectively, so that the guard bandwidth is 250 Hz, and the remaining 448 subcarriers are divided into 64 groups with four pilot subcarriers and three data subcarriers in each group. Frequency diversity technology is applied against channel multi-path fading. The OFDM signal bandwidth is 4 kHz and the modulation is QPSK. The signal parameters are summarized in Table 2.

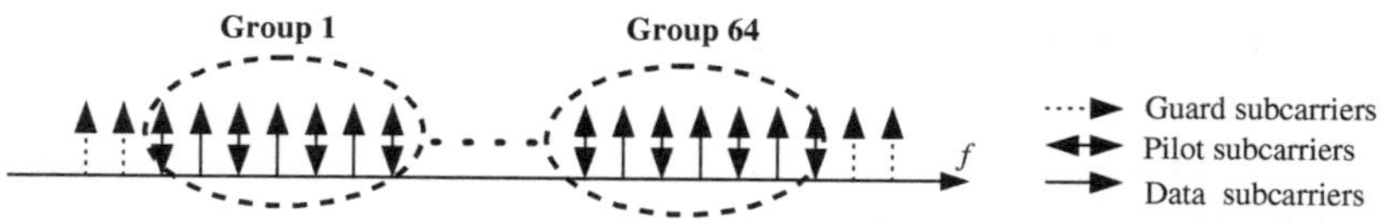

Fig. 2 The arrangement of subcarriers

Table 2 OFDM signal parameters

K	Guard subcarriers K_g	Pilot subcarriers K_p	Data subcarriers K_a	Bit rates $2K_a/(T + T_g)$	Bit rates 2 diversity	Bit rates 4 diversity
512	64	256	192	2400 bps	1200 bps	600 bps

3.2 *Doppler Estimation and Compensation*

In OFDM, the method of Doppler estimation and compensation is similar to DSSS. PN sequence is chosen as the pilot data which are used to estimate the fine symbol timing and Doppler spread, as well as the traditional channel response function. The received signal is first transformed to a frequency domain by FFT. Once the output of the cross-correlation pilot data with known PN sequence reaches maximum, the symbol timing and Doppler estimation are achieved simultaneously.

4 Time Reversal Process

4.1 *Signal Frame Format*

Figure 3 shows the transmit signal frame format. The LFM signal is inserted at the beginning of each information symbol, which is used for symbol synchronization and channel probe for passive time reversal. Blank time remains between the probe and information symbol. The bandwidth of the LFM signal spans the full bandwidth of the information symbol.

4.2 *Receiver Process*

The time reversal approach that we are developing in this system avoids the explicit recovery of the channel and its subsequent equalization. Figure 4 is the time reversal process flow diagram, where we use the frequency domain equivalents of all signals. At the *i*th element of the receiver array, the channel impulse response function is $h_i(t)$ and its Fourier transform is $H_i(w)$. Via the matching filter, the output of the received probe signal is $|P(w)|^2 H_i(w)$. The received information signal is $S(w)H_i(w)$. The output of a passive time reversal process is obtained by multiplying the received information signal and the conjugation of $|P(w)|^2 H_i(w)$

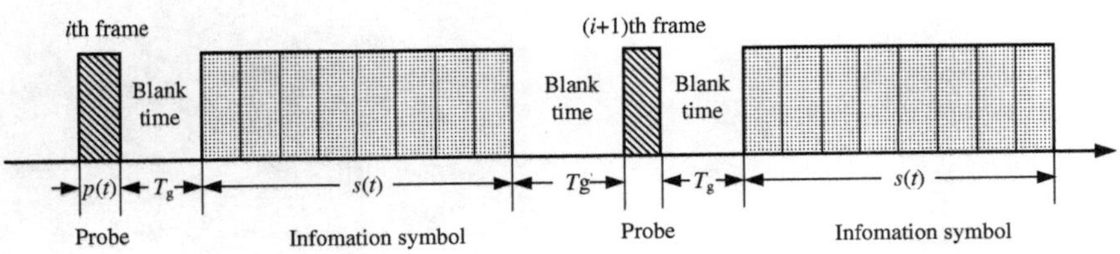

Fig. 3 Signal frame format

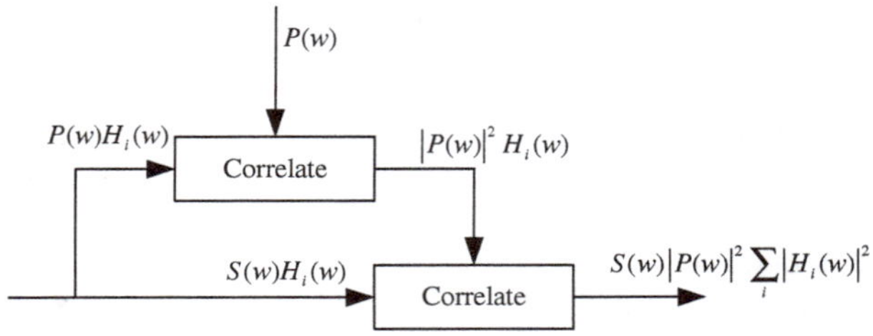

Fig. 4 Time reversal process flow diagram

and summing over all elements, which results in $S(w)|P(w)|^2\sum_i |H_i(w)|^2$. In the bandwidth of the signal, $|P(w)|^2\sum_i |H_i(w)|^2$ is nearly constant, so the time reversal process can partly reduce the multi-path distortion.

5 Modem Equipment

Modem DSP is TI's fixed-point processor TMS320C6455 running at 1.2 GHz and obtaining a convolutional code coprocessor. The communication signal bandwidth is 6–10 kHz. The topside modem is equipped with a 4-element hydrophone array and a transceiver transducer. The array element spacing is 0.75 m, which is equal to 4 times the wavelength of a signal at a center frequency of 8 kHz. The maximum transmit source level is 183 dB (re 1 μPa @ 1 m). The communication link between the acoustic modem and the PC is RS422, and the communication link between the acoustic modem and the AUV is CAN2.0. Figure 5 shows the topside modem with a 4-element hydrophone array. Figure 6 shows the transducer mounted onto an AUV.

Fig. 5 Topside modem with a 4-element hydrophone array

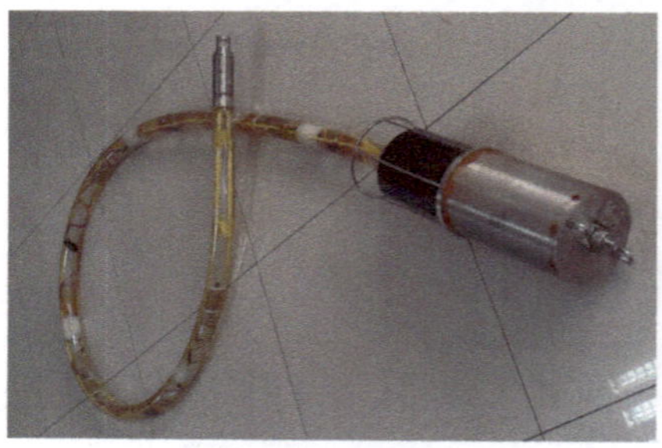

Fig. 6 Transducer mounted onto an AUV

6 Experiment Description and Results

The AUV acoustic communication experiments were performed at Qiandao Lake in August and in South China Sea in September 2013. The body of water that the AUV was deployed into has an average depth of approximately 30 m. The AUV was programmed to operate at a fixed depth of 20 m below the water surface. The relative speed between the AUV and the mother ship is up to 2.5 m/s. The data that was transmitted between the topside modem and the AUV modem included the following: AUV calculated latitude, AUV calculated longitude, depth below the surface, height on the bottom, etc. Currently, the size of the data packet sent from the topside modem to the AUV is 40 bytes.

In various tests, communications with the AUV were achieved at ranges up to approximately 1500 m, and the data packet was sent back stably from the AUV to the topside modem every 20 s. Occasionally, the data packet CRC would cause an error when the AUV made a turn with the AUV depth changing rapidly. Figure 7 shows the AUV operator display interface, where the lines show AUV position information for tracking obtained from the acoustic modem.

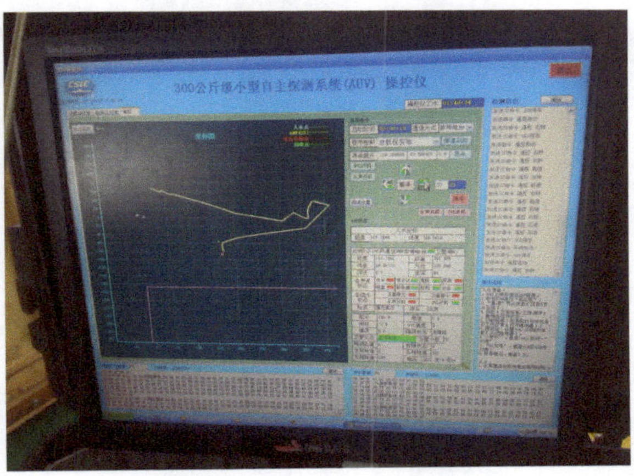

Fig. 7 AUV operator display interface

7 Summary and Future Work

This paper designed an acoustic communication system for an AUV, which integrates DSSS and OFDM. Doppler caused by the AUV moving is estimated and compensated accurately. The time reversal process is used to restrain multi-path interference. Through testing in the lake and sea, the results show that this acoustic communication system can provide near real-time AUV position information back for the operator, as well as important information about the status of the mission and the AUV. Finally, the future work should include decreasing the power of the modem, using FPGA to achieve Doppler search, selection of the communication mode and speed adaptively.

Acknowledgments This work is supported by the National "863" Program of China (No. 2011AA09 A107).

References

1. Stojanovic, M., Proakis, J., Rice, J., Green, D.: Spread spectrum for underwater acoustic telemetry. In: OCEANS'98 Conference Proceedings, vol. 2, pp. 650–654 (1998)
2. Stojanovic, M., Freitag, L.: Multichannel detection for wideband underwater acoustic CDMA communications. IEEE J. Oceanic Eng. **31**(3), 685–695 (2006)
3. Blackmon, F., Sozer, E., Stojanovic, M., Proakis, J.: Performance comparison of RAKE and hypothesis feedback direct sequence spread spectrum techniques for underwater communication applications. In: OCEANS-Conference, vol 1, pp. 594–603 (2002)
4. Li, B., Zhou, S., Stojanovic, M., Freitag, L., Willett, P.: Multicarrier communication over underwater acoustic channels with nonuniform Doppler shifts. Oceanic Eng. **33**(2), 198–209 (2008)
5. Gomes, J., Silva, A., Jesus, S.: Experimental assessment of time-reversed OFDM underwater communications. J. Acoust. Soc. Am. **123**(5), 3891 (2008)
6. Kang, T., Song, H., Hodgkiss, W.S.: OFDM underwater acoustic communications in KAM08. The 4th ACM international workshop on underwater networks (2009)

Research of Axis Mismatches Between Pairs of Sensitive Elements of Underwater Acoustic Velocity Gradient Sensors

Xinyi Sun, Desen Yang, Lianjin Hong, Shengguo Shi and Hongkun Zhou

Abstract Underwater acoustic velocity gradient sensors can be combined with one pressure sensor and six biaxial vector sensors as sensitive elements, and axis mismatches between pairs of sensitive elements are studied in detail in this paper. The formulations of measurement errors of particle velocities and velocity gradients are measured by underwater acoustic velocity gradient sensors and are derived while the axis mismatch exists, then the influences of measurement errors of underwater acoustic velocity gradient sensors by axial mismatches are analyzed.

Keywords Underwater acoustics · Velocity gradient sensor · Axis mismatch

1 Introduction

Underwater acoustic velocity gradient sensors arise from the second-order approximation of the Taylor series expansion of the pressure field. They are a type of dyadic sensor, and they can measure scalar pressure, vector particle velocity, and tensor velocity gradient, etc. [1]. Underwater acoustic velocity gradient sensors have three types of directivities: monopole directivity, dipole directivity, and quadrupole directivities [2], and quadrupole directivities contain both longitudinal quadrupole directivity and lateral quadrupole directivity. An underwater acoustic velocity gradient sensor can produce a beamwidth of 65° compared to 105° for a vector sensor. Therefore, researchers focus on high-order directional

X. Sun · D. Yang · L. Hong · S. Shi · H. Zhou
College of Underwater Acoustic Engineering,
Harbin Engineering University, Harbin 150001, China

D. Yang · L. Hong · S. Shi
Science and Technology on Underwater Acoustic Laboratory,
Harbin Engineering University, Harbin 150001, China

X. Sun (✉)
Naval Academy of Armament, Shanghai 200436, China
e-mail: sunxinyi@hrbeu.edu.cn

L. Zhou et al. (eds.), *Underwater Acoustics and Ocean Dynamics*
DOI 10.1007/978-981-10-2422-1_10

sensors and consider putting them into arrays to improve the performance of the arrays [3, 4].

Underwater acoustic velocity gradient sensors can be combined with one pressure sensor and six biaxial vector sensors, and the structure for underwater acoustic velocity gradient sensors is given in this paper. With this type of design, particle velocity and velocity gradient will be measured approximately. Then the formulations of measurement errors of underwater acoustic velocity gradient sensors in theoretically ideal conditions are derived. Whereas, in practice, axis mismatches exist between pairs of vector sensors, and the formulations of measurement errors as axis mismatches can be obtained. Finally, the formulations are analyzed by figures in this paper, and the relationship among measurement errors, axis mismatches, azimuth angles of sources, and the frequency range of underwater acoustic velocity gradient sensors are discussed.

2 Taylor Series Expansion of an Acoustic Pressure Field

An nth-order Taylor series expansion of an acoustic pressure field contained within a finite region $(x - x_0, y - y_0, z - z_0)$ about the point (x_0, y_0, z_0) can be expressed as

$$p(\boldsymbol{r}) = p(\boldsymbol{r}_0) + \sum_{n=1}^{\infty} \frac{1}{n!}[(\boldsymbol{r} - \boldsymbol{r}_0)\nabla]^n p(\boldsymbol{r}_0) \tag{1}$$

where $\boldsymbol{r} = [x\ y\ z]$ and $\boldsymbol{r}_0 = [x_0\ y_0\ z_0]$. The spatial gradient of pressure can be related to acoustic particle velocity via the linearized momentum equations

$$\frac{\partial p(\boldsymbol{r}_0)}{\partial x_0} = -j\omega\rho_0 u_x, \frac{\partial p(\boldsymbol{r}_0)}{\partial y_0} = -j\omega\rho_0 u_y, \frac{\partial p(\boldsymbol{r}_0)}{\partial z_0} = -j\omega\rho_0 u_z \tag{2}$$

where ρ_0 is the ambient density of the surrounding fluid, and u_x, u_y, and u_z are the orthogonal components of the acoustic particle velocity vector. By substituting Eq. (2) into Eq. (1), taking the appropriate partial derivatives, and expanding the series to the second order (n = 2), Eq. (1) can be rewritten as

$$p(\boldsymbol{r}) \approx p(\boldsymbol{r}_0) - j\omega\rho_0[\boldsymbol{r} - \boldsymbol{r}_0]\begin{bmatrix} u_x \\ u_y \\ u_z \end{bmatrix} - \frac{1}{2}j\omega\rho_0[\boldsymbol{r} - \boldsymbol{r}_0]\begin{bmatrix} \frac{\partial u_x}{\partial x_0} & \frac{\partial u_x}{\partial y_0} & \frac{\partial u_x}{\partial z_0} \\ \frac{\partial u_y}{\partial x_0} & \frac{\partial u_y}{\partial y_0} & \frac{\partial u_y}{\partial z_0} \\ \frac{\partial u_z}{\partial x_0} & \frac{\partial u_z}{\partial y_0} & \frac{\partial u_z}{\partial z_0} \end{bmatrix}[\boldsymbol{r} - \boldsymbol{r}_0]^{\mathrm{T}} \tag{3}$$

Notice that the pressure is a scalar quantity, equivalent to a tensor of rank zero; the velocity is a vector quantity, a first-rank tensor combined with three orthogonal components; the velocity gradient is a second-rank tensor combined with nine components. The propagating plane wave field is irrotational and, hence, the curl of the velocity vector is zero [5]

$$\begin{cases} \frac{\partial u_x}{\partial y_0} = \frac{\partial u_y}{\partial x_0} \\ \frac{\partial u_x}{\partial z_0} = \frac{\partial u_z}{\partial x_0} \\ \frac{\partial u_y}{\partial z_0} = \frac{\partial u_z}{\partial y_0} \end{cases} \quad (4)$$

Then Eq. (3) can be simplified as

$$p(\boldsymbol{r}) \approx p(\boldsymbol{r}_0) - j\omega\rho_0[\boldsymbol{r} - \boldsymbol{r}_0]\begin{bmatrix} u_x \\ u_y \\ u_z \end{bmatrix} - \frac{1}{2}j\omega\rho_0[\boldsymbol{r} - \boldsymbol{r}_0]\begin{bmatrix} \frac{\partial u_x}{\partial x_0} & \frac{\partial u_x}{\partial y_0} & \frac{\partial u_x}{\partial z_0} \\ \frac{\partial u_x}{\partial y_0} & \frac{\partial u_y}{\partial y_0} & \frac{\partial u_y}{\partial z_0} \\ \frac{\partial u_x}{\partial z_0} & \frac{\partial u_y}{\partial z_0} & \frac{\partial u_z}{\partial z_0} \end{bmatrix}[\boldsymbol{r} - \boldsymbol{r}_0]^{\mathrm{T}} \quad (5)$$

Notice that the second-order terms above can be divided into two types: the pure partial derivatives of the velocity and the mixed partial derivatives of the velocity corresponding to the diagonal terms and off-diagonal, respectively. An acoustic velocity gradient sensor should measure pressure, particle velocity, and velocity gradient (ten components in plane wave field) at one point $\boldsymbol{r}_0$.

3 Structure of Velocity Gradient Sensors

A velocity gradient sensor is combined with one pressure sensor and three pairs of biaxial vector sensors as sensitive elements, as shown in Fig. 1. The pressure sensor is in the center of the velocity gradient sensor, three pairs of collinear vector sensors are laid on the coordinate axes (x, y, z), and each pair is oppositely positioned relative to the center. The predetermined distance between each pair of vector sensors is L, and the sensitive axes are shown by the arrows in Fig. 1.

For this type of design, the acoustic pressure p can be measured directly, and the particle velocity components and velocity gradient components are obtained by principles of mean value and finite-difference approximation, respectively. For example

$$\begin{cases} u_x \approx \frac{u_{2x} + u_{1x}}{2} \\ \frac{\partial u_x}{\partial x} \approx \frac{u_{2x} - u_{1x}}{L} \\ \frac{\partial u_y}{\partial x} \approx \frac{u_{2y} - u_{1y}}{L} \end{cases} \quad (6)$$

4 A Velocity Gradient Sensor in a Plane Wave Field

A velocity gradient sensor is simplified to a single axis and the plane wave field is simplified to two dimensions, as shown in Fig. 2. The analysis of the other two axes of the velocity gradient sensor would be analogous. A pair of biaxial vector sensors

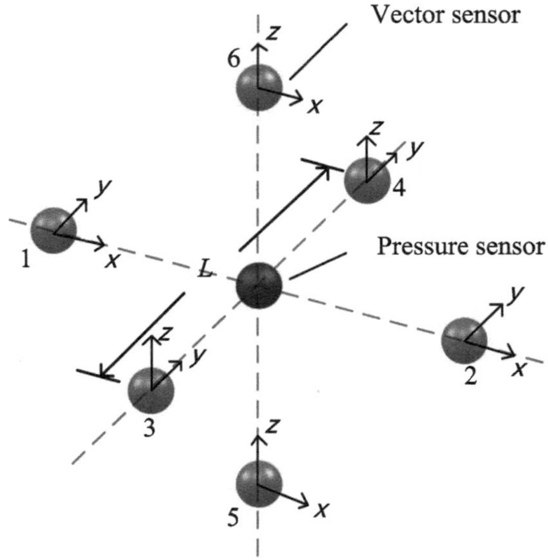

Fig. 1 The structure sketch of a velocity gradient sensor

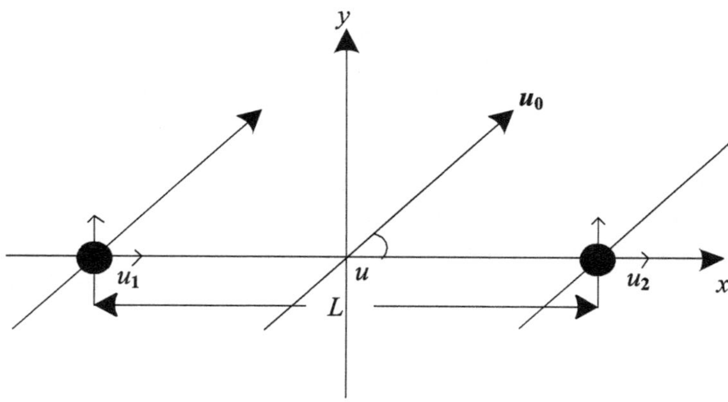

Fig. 2 A velocity gradient sensor in a plane wave field

are laid in u_1 and u_2 positions respectively, and the sensor spacing is L. The azimuth angle between the x-axis and the propagation direction of the plane wave is θ. The sensitive axes of the vector sensors and the direction of the plane wave are shown by the arrows.

The components of the velocity at the points u_1 and u_2 are

$$\begin{cases} u_{1x} = \cos\theta u_0 e^{j\left[\omega t - k\left(x - \frac{L}{2}\right)\cos\theta - ky\sin\theta\right]} \\ u_{1y} = \sin\theta u_0 e^{j\left[\omega t - k\left(x - \frac{L}{2}\right)\cos\theta - ky\sin\theta\right]} \\ u_{2x} = \cos\theta u_0 e^{j\left[\omega t - k\left(x + \frac{L}{2}\right)\cos\theta - ky\sin\theta\right]} \\ u_{2y} = \sin\theta u_0 e^{j\left[\omega t - k\left(x + \frac{L}{2}\right)\cos\theta - ky\sin\theta\right]} \end{cases} \tag{7}$$

where u_0 is the amplitude of the plane wave, ω is angular frequency ($\omega = 2\pi f$, f is the source frequency), and k is the wave number ($k = \omega/c$, c = 1500 m/s). As shown in Fig. 2, the pure partial derivative of the velocity $\frac{\partial u_x}{\partial x}$ can be written as

$$\frac{\partial u_x}{\partial x} = -jk\cos^2\theta u_0 e^{j[\omega t - kx\cos\theta - ky\sin\theta]} \tag{8}$$

Then the mixed partial derivative of the velocity $\frac{\partial u_y}{\partial x}$ can be written as

$$\frac{\partial u_y}{\partial x} = -jk\cos\theta\sin\theta u_0 e^{j[\omega t - kx\cos\theta - ky\sin\theta]} \tag{9}$$

From Eqs. (6) to (9), the measurement errors of the particle velocity and velocity gradient can be expressed as

$$\begin{cases} \varepsilon_{u_x} = 20\lg\left|\frac{(u_{1x} + u_{2x})/2}{u_x}\right| = 20\lg\left|\cos\left(\frac{2\pi fL\cos\theta}{c}\right)\right| \\ \varepsilon_{\frac{\partial u_x}{\partial x}} = 20\lg\left|\frac{(u_{2x} - u_{1x})}{L} \Big/ \frac{\partial u_x}{\partial x}\right| = 20\lg\left|\frac{\sin(2\pi fL\cos\theta/c)}{2\pi fL\cos\theta/c}\right| \\ \varepsilon_{\frac{\partial u_y}{\partial x}} = 20\lg\left|\frac{\left(u_{2y} - u_{1y}\right)}{L} \Big/ \frac{\partial u_y}{\partial x}\right| = 20\lg\left|\frac{\sin(2\pi fL\cos\theta/c)}{2\pi fL\cos\theta/c}\right| \end{cases} \tag{10}$$

5 Axis Mismatches Between Pairs of Vector Sensors

Limited to the technic and elastic suspension of vector sensors, there will be an included angle (θ_m) which exists between the same sensitive axis of the vector sensors in practice, as shown in Fig. 3.

Due to the included angle θ_m, the components (u_{1x}, u_{1y}) of the velocity of the No. 1 vector sensor will be changed and expressed as

$$\begin{cases} u_{1x} = \cos(\theta - \theta_m) u_0 e^{j\left[\omega t - k\left(x - \frac{L}{2}\right)\cos\theta - ky\sin\theta\right]} \\ u_{1y} = \sin(\theta - \theta_m) u_0 e^{j\left[\omega t - k\left(x - \frac{L}{2}\right)\cos\theta - ky\sin\theta\right]} \end{cases} \tag{11}$$

Then the measurement errors of the particle velocity and velocity gradient are changed and expressed as

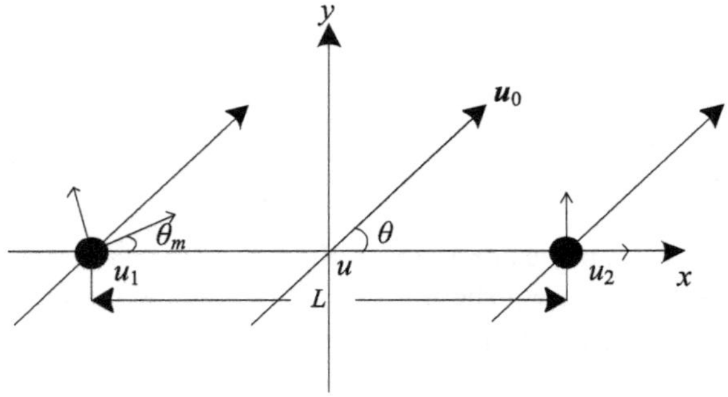

Fig. 3 Axis mismatches between pairs of vector sensors

$$\begin{cases} \varepsilon_{u_x} = 20\lg\left|\frac{(u_{2x}+u_{1x})/2}{u_x}\right| = 20\lg\left|\frac{\cos(\theta-\theta_m)e^{j\pi f\frac{L}{c}\cos\theta}+\cos\theta e^{-j\pi f\frac{L}{c}\cos\theta}}{2\cos\theta}\right| \\ \varepsilon_{\frac{\partial u_x}{\partial x}} = 20\lg\left|\frac{(u_{2x}-u_{1x})}{L}\frac{\partial u_x}{\partial x}\right| = 20\lg\left|\frac{\cos\theta e^{-j\pi f\frac{L}{c}\cos\theta}-\cos(\theta-\theta_m)e^{j\pi f\frac{L}{c}\cos\theta}}{-2j\pi fL\cos^2\theta/c}\right| \\ \varepsilon_{\frac{\partial u_y}{\partial x}} = 20\lg\left|\frac{(u_{2y}-u_{1y})}{L}\frac{\partial u_y}{\partial x}\right| = 20\lg\left|\frac{\sin\theta e^{-j\pi f\frac{L}{c}\cos\theta}-\sin(\theta-\theta_m)e^{j\pi f\frac{L}{c}\cos\theta}}{-2j\pi fL\cos\theta\sin\theta/c}\right| \end{cases} \tag{12}$$

Assume that the included angle $\theta_m = 5°$, and Eq. (12) is drawn as contour maps and the 3-D surface plots of measurement errors ε versus *fL* and azimuth angle θ, as shown in Fig. 4.

Figure 4 shows that when the azimuth angle θ is near $\pm\pi/2$, the measurement errors of the particle velocity u_x and velocity gradient $\frac{\partial u_x}{\partial x}$ are huge, and the measurement error of the velocity gradient $\frac{\partial u_y}{\partial x}$ is huge when the azimuth angle θ is near 0, $\pm\pi/2$ and $\pm\pi$.

Therefore, for a fixed permissible error, it cannot be satisfied on every azimuth angle θ. There should be a region to meet the permissible error, and the region should be a beamwidth. When permissible errors of the particle velocity and velocity gradient are within 1 dB in the region of the beamwidth, then for particle velocity u_x, the range of *fL* is

$$0 \le fL \le 218 \tag{13}$$

For velocity gradient $\frac{\partial u_x}{\partial x}$, the range of *fL* is

$$29 \le fL \le 380 \tag{14}$$

For velocity gradient $\frac{\partial u_y}{\partial x}$, the range of *fL* is

$$149 \le fL \le 244 \tag{15}$$

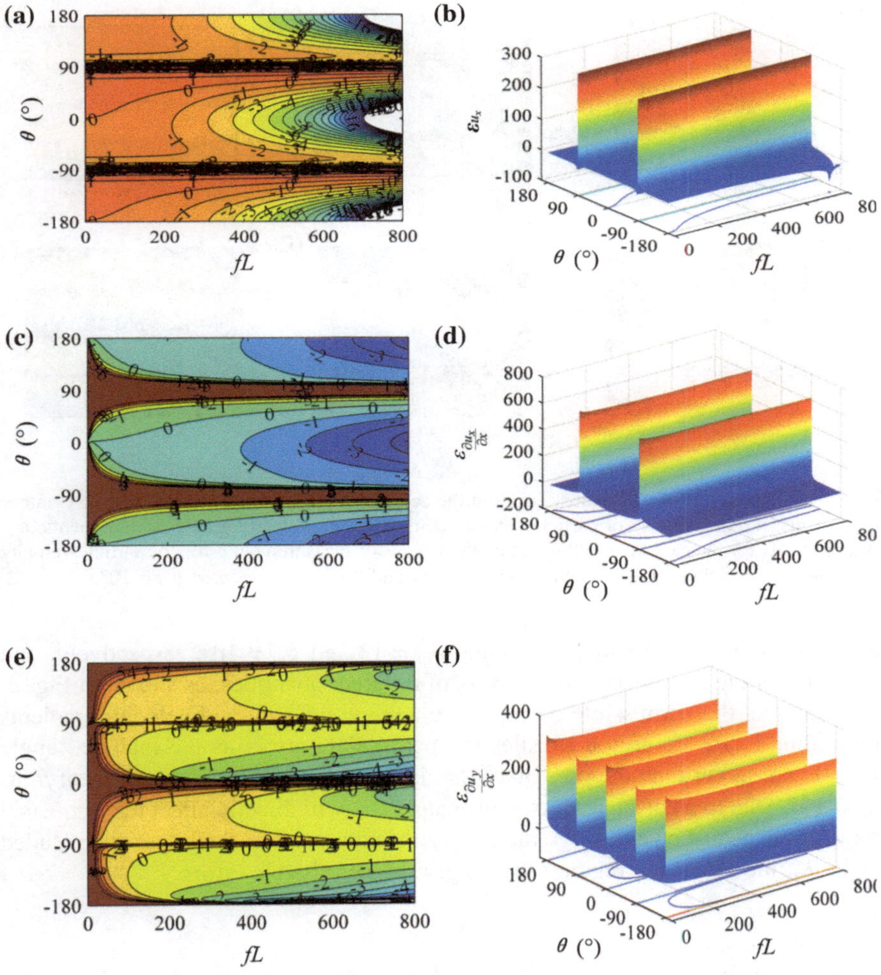

Fig. 4 Measurement errors while axis mismatches exist ($\theta_m = 5°$). **a**, **c**, **e** Contour map of measurement errors of particle velocity component u_x, velocity gradient components $\partial u_x/\partial x$ and $\partial u_y/\partial x$, respectively; **b**, **d**, **f** 3-D surface plot of measurement errors of particle velocity component u_x, velocity gradient components $\partial u_x/\partial x$ and $\partial u_y/\partial x$, respectively

From Eqs. (13) to (15), the intersection of fL is

$$149 \le fL \le 218 \tag{16}$$

Therefore, when spacing L is fixed, the frequency range of the acoustic velocity gradient sensors is fixed. Equations (13)–(15) show that for more quantities measured approximately, the smaller will be the intersection of fL, and moreover, the intersection of fL will be smaller or disappear as the included angle θ_m becomes

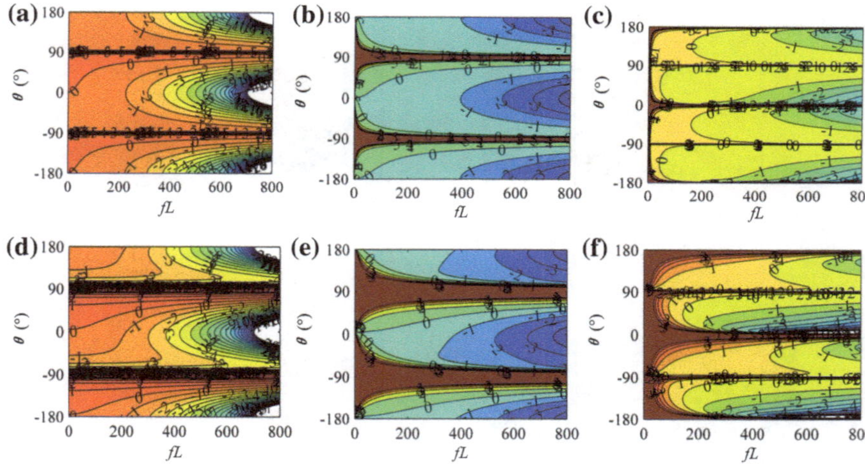

Fig. 5 Comparison of measurement errors in the cases of $\theta_m = 1°$ and $\theta_m = 10°$. **a**, **b**, **c** Contour maps of measurement errors of particle velocity component u_x, velocity gradient components $\partial u_x/\partial x$ and $\partial u_y/\partial x$ for the case $\theta_m = 1°$; **d**, **e**, **f** contour maps of measurement errors of particle velocity component u_x, velocity gradient components $\partial u_x/\partial x$ and $\partial u_y/\partial x$ for the case $\theta_m = 10°$

larger. Assume that the included angle $\theta_m = 1°$ and $\theta_m = 10°$, respectively, and then compare to the measurement errors of different quantities, as shown in Fig. 5. For example, the permissible errors of the particle velocity and velocity gradients are within 1 dB, the region satisfies the permissible errors as the included angle $\theta_m = 1°$ are larger than the case that $\theta_m = 10°$, and there is no intersection of fL as $\theta_m = 10°$. Due to a constant L, axis mismatch (θ_m) will directly affect the frequency range of the acoustic velocity gradient sensors and moreover, the larger the included angle θ_m, the larger the region over ranges the permissible errors.

6 Conclusion

Velocity gradient sensors can be combined with one pressure sensor and six biaxial vector sensors, and they can measure ten quantities in a plane wave field, including pressure, three components of particle velocity and six components of velocity gradients. The formulations of measurement errors of acoustic velocity gradient sensors are obtained when axis mismatches exist, and the conclusions are summarized through the formulations as follows.

The performance of acoustic velocity gradient sensors will be strongly affected by axis mismatches between pairs of sensitive elements. Due to axis mismatches, the acoustic velocity gradient sensors have higher measurement errors of particle velocity and velocity gradient in some regions, such as the azimuth angle θ is near 0, $\pm\pi/2$ and $\pm\pi$. When axis mismatches become larger, the region of high measurement errors will increase. For the fixed permissible error and spacing L, the frequency range of the acoustic velocity gradient sensors is directly affected by axis mismatches between pairs of collinear vector sensors. The frequency range of acoustic velocity gradient sensors will be narrowed with increasing axis mismatches.

Acknowledgments This work was supported by the National Natural Science Foundation of China (grant Nos. 11074057, 11074058 and 11204049), Program for Changjiang Scholars and Innovative Team in University (No. IRT1228), and the Specialized Research Fund for the Doctoral Program of Higher Education (grant Nos. 20122304120023 and 20122304120011). This paper is funded by the International Exchange Program of Harbin Engineering University for Innovation-Oriented Talents Cultivation.

References

1. Silvia, M.T.A.: Theoretical and experimental investigation of acoustic dyadic sensors. SITTEL Technical Report No. TP-4. SITTEL Corporation, Ojai, California (2001)
2. Yang, D.S., Sun, X.Y., Hong, L.J., et al.: The velocity gradient sensor based on the vector hydrophone. J. Harbin Eng. Univ. **34**(1), 7–14 (2013)
3. Cox, H., Lai, H.: Performance of line arrays of vector and higher order sensors. In: Conference Record of the Forty-First Asilomar Conference on Signals, Systems and Computers, 2007 (ACSSC 2007), pp. 1231–1236. IEEE (2007)
4. Song, Y., Wong, K.T.: Closed-form direction finding using collocated but orthogonally oriented higher order acoustic sensors. IEEE Sens. J. **12**(8), 2604–2608 (2012)
5. Cray, B.A., Victor, M.E., Albert, H.N.: Highly directional acoustic receivers. J. Acoust. Soc. Am. **113**(3), 1526–1532 (2003)

A Passive Fathometer Technique for Bottom Profiling Using Ambient Noise

Junghun Kim, Jee Woong Choi and Jungyul Na

Abstract A passive fathometer technique using ambient noise can be used for estimating water depth and the thickness of the sub-bottom layer. This technique uses the correlation of sea surface-generated noise estimated for signals received by a vertical hydrophone array via a beamforming process. In this paper, the nested vertical line array (N-VLA) was used for effective broadband beamforming, and the beamforming result for each sub-array was applied to the passive fathometer technique. The estimated results are presented here and compared to the sediment information obtained by chirp sonar profiles and coring analysis.

Keywords Passive fathometer technique · Nested vertical line array · Ambient noise

1 Introduction

This paper presents the estimated results for the water depth and the thickness of the sub-bottom layer using surface-generated ambient noise. There have been several studies related to a passive fathometer technique using ambient noise. Recently, Siderius et al. [1] applied the conventional and adaptive beamforming methods to the passive fathometer technique, and showed that the estimate predicted by the adaptive process had a better performance than that of a conventional process. In this paper, both techniques are also applied to the ambient noise data, acquired by a nested vertical line array (N-VLA) deployed off the coast of Donghae (East Sea),

J. Kim (✉) · J.W. Choi · J. Na
Department of Marine Sciences and Convergent Technology,
Hanyang University, Ansan, Korea
e-mail: kimjh0927@onestx.com

J.W. Choi
e-mail: choijw@hanyang.ac.kr

J. Na
e-mail: najy0252@hanyang.ac.kr

L. Zhou et al. (eds.), *Underwater Acoustics and Ocean Dynamics*
DOI 10.1007/978-981-10-2422-1_11

Korea. Beamforming for each sub-array was performed, a proper frequency range for each sub-band was chosen, in which no grating lobe appeared, the mainlobe width was sufficiently narrow, and the result was then applied to the passive fathometer technique.

2 Acoustic Measurements

Acoustic measurements were made on July 13, 2009. The experimental site was about 10 km off the eastern coast of Korea (Fig. 1a). The nominal water depth was about 150 m. The ambient noise was measured by an N-VLA called POEMS, covering the water column between 92.5 and 107.5 m. The POEMS consists of a total of 24 hydrophone channels and four sub-bands (nested arrays) (Fig. 1b). During the acoustic measurements, sound speed profiles were monitored using an expendable bathy thermograph (XBT). The bottom topography was obtained with a multibeam echo sounder. The results determined by Korea Institute of Geoscience and Mineral Resources (KIGAM) showed that from the core analysis and chirp sonar profiles there was a sediment layering in the experimental area [2]: the first sub-bottom layer exists 1 m below the water-sediment interface and the second sub-bottom layer is between 15 and 20 m.

3 Results

Figure 2a shows the results of the passive fathometer technique obtained using conventional beamforming (CBF) and adaptive MVDR methods. Note that the amplitudes were normalized to the maximum value of the outputs. Figure 2b shows

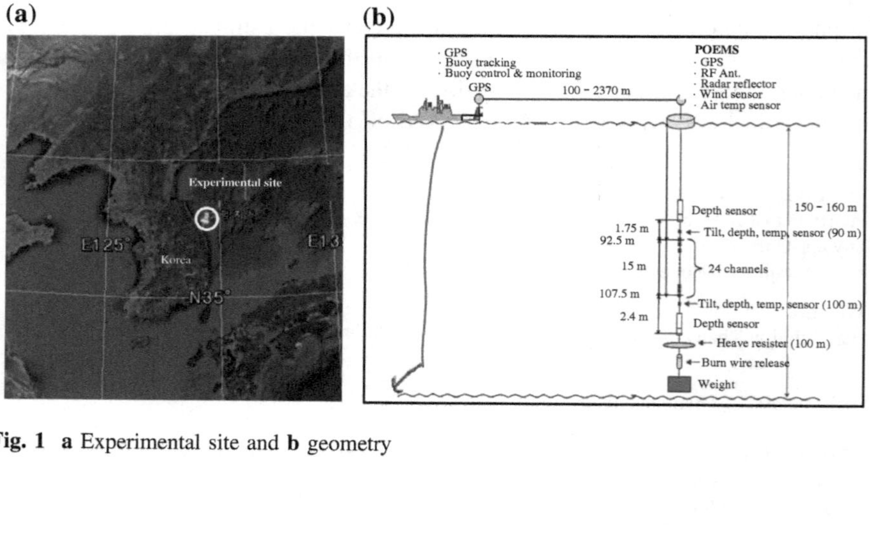

Fig. 1 **a** Experimental site and **b** geometry

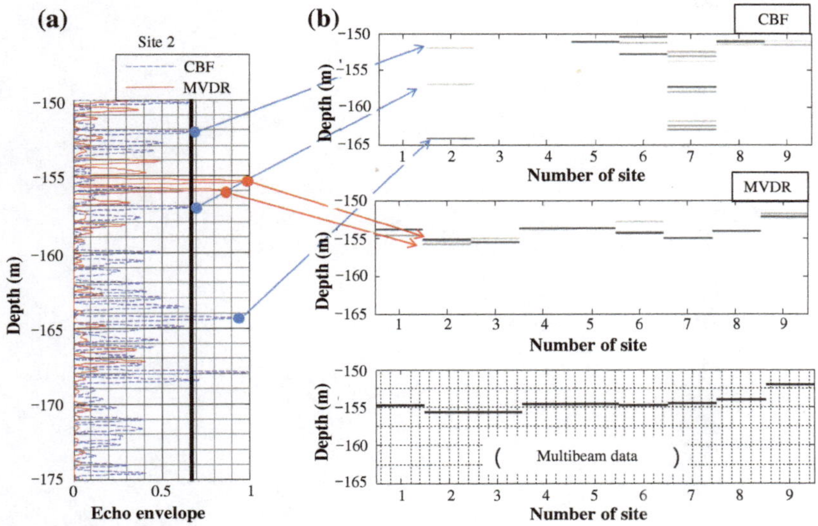

Fig. 2 **a** Results of passive fathometer technique obtained using CBF and MVDR methods at location 2#; **b** the water depth estimates for 9 different locations

the estimated water depths for nine different measurement locations. The top and middle figures show the results estimated by CBF and the adaptive MVDR beamforming methods, respectively; the bottom shows the depth of the water–sediment interface estimated by a multibeam echo sounder. The results of the MVDR beamforming method show a much better performance than those of the CBF method, resolving the O (1 m)-thick surficial sediment layer. However, the second interface was not successfully resolved from the passive fathometer technique in the case of even MVDR outputs. It is likely that the second layer was not resolved because of the relatively lower source level. It was reported that as wind speed increases, the performance has higher efficiency [3]. In our case, the wind speed was less than 7 m/s during the period of acoustic measurements, which means that the source level was relatively low.

Acknowledgments This work was supported by the Agency for Defense Development, Korea (UD130007DD).

References

1. Siderius, M., Song, H., Gerstoft, P., Hodgkiss, W.S., Hursky, P., Harrison, C.: Adaptive passive fathometer processing. J. Acoust. Soc. Am. **127**(4), 2193–2200 (2010)
2. Korea Institute of Geoscience and Mineral Resources. Research Report: NP2007-010 (2007)
3. Means, S.L., Siderius, M.: Effects of sea-surface conditions on passive fathometry and bottom characterization. J. Acoust. Soc. Am. **126**(5), 2234–2241 (2009)

Target Motion Parameter Estimation for LOFARgrams Based on Waveguide Invariants

Jian Li, Guangjie Han, Defu Zhou, Kai Tang and Qingbang Han

Abstract To estimate the motion parameters of a wideband moving target which is passing by the closest point of approach (CPA) of the receiving linear array, a theoretical method based on detection and analysis parabolic striations (the striations can be extracted using an image processing technique such as the HOUGH transform) of the low-frequency analyzing and recording gram data (LOFARgram) is proposed. While a traditional approach uses one sensor with one LOFARgram, the method proposed here needs at least two LOFARgrams (sensors) to calculate the motion parameters. First, the theoretical equations for calculating the waveguide invariant (WI) target course angle, velocity and range from the LOFARgram are derived. Then, the method is employed to process the LOFARgrams which are obtained from an acoustic toolbox simulation. The parameters estimation results show the validity of the method.

Keywords Passive detection · Closest point of approach · Acoustic toolbox

J. Li (✉) · G. Han · K. Tang
College of Internet of Things Engineering, Hohai University, Changzhou 213022, China
e-mail: jian263@sina.com

J. Li
State Key Laboratory of Acoustics, Institute of Acoustics, CAS, Beijing 100190, China

D. Zhou
Key Laboratory of Technology and Application for Safeguarding of Marine Rights and Interests, State Oceanic Administration, Guangzhou 510310, China

D. Zhou
South China Sea Marine Engineering Surveying Center, South China Sea Branch, State Oceanic Administration, Guangzhou 510310, China

Q. Han
Changzhou Key Laboratory of Sensor Networks and Environmental Sensing, Hohai University, 213022 Changzhou, China

L. Zhou et al. (eds.), *Underwater Acoustics and Ocean Dynamics*
DOI 10.1007/978-981-10-2422-1_12

1 Introduction

Underwater target detection is still an important but difficult problem in the domain of underwater acoustic research. Detection sonar mainly uses the passive mode for the purpose of concealment; the sonar system estimates the distance, bearing and velocity of the target through acquisition and processing target radiated noise. Passive detection methods primarily include: (1) The three-point ranging method, based on the spherical or cylindrical wave front theory; it can estimate the bearing and range of the target; the accuracy of this method depends on the delay estimation accuracy, the target distance, the target direction, the array aperture, the array installation accuracy [1, 2], and so on. (2) The TMA (target motion analysis) method, based on the goal bearing-time analysis; this includes the bearing TMA, the frequency-bearing TMA, the multipath TMA, and other methods [3]. (3) The passive multipath ranging method, based on the triangulation principle [4]; the ranging accuracy of this method depends on the prediction of the multipath structure. (4) Beamforming, which includes conventional beamforming, MVDR beamforming, near field focus beamforming [5], and so on. (5) Model-based ranging methods, based on the normal mode description of the sound field; these methods include MFP (matched field processing), MMP (matched mode processing), and PTRM (passive time reversal mirror) [3, 6, 7]. These methods must establish the channel model based on the true environmental parameters in order to calculate the replica sound field. Performance comparisons of the above methods are shown in Table 1. (6) Vector method, which uses a vector sensor to measure the scalar pressure and the particle velocity (P, V_x, V_y, V_z) at the same point and at the same time [8], so we have more choices when we do BF or MFP processing. (7) Waveguide invariant (WI) method [9, 10]. WI explains the interference stripes on the range-frequency plane, the range information can be derived through analyzing the received broadband noise of the target.

The methods mentioned above are listed in Table 1.

The process is: first, methods based on ray geometry theory ignore the influence of the underwater acoustic channel, such as method No. 4; second, methods based on an underwater acoustic model are difficult in finding optimal parameters to fit the real environment, such as method No. 5; third, the methods consider the channel effect, but independent of the environment parameters proposed, such as method No. 7.

In the situation where the target distance changes with time, if we apply a low-frequency analysis and record (LOFAR) analysis to the received broadband noise signal, the emerged regular stripes will display on the received LOFAR figure. This is due to the sound field's normal wave interference, which changes with the distance; this phenomenon is known as the WI, denoted as β. If the distance of the target experiences a far-near-far variation, the interference stripes will resemble a parabola. In Ref. [10], a method of passive ranging based on a WI was proposed; however, the method has the hypothesis of a known target velocity. As a supplement, here we put forward an improved method.

Table 1 List of different passive detection methods

No.	Method	Purpose	Working distance and influence factors
1	Three-point ranging	Bearing and ranging	Near; delay estimation accuracy, the array aperture and installation accuracy, target distance, target direction, etc.
2	TMA	Target position based on bearing estimation	Middle and far; the measurement accuracy of frequency and direction
3	Passive multipath ranging	Ranging	Middle and near; the multipath structure and signal bandwidth
4	Beamforming (CBF, MVDR, focused BF)	Bearing and ranging	Near; the array elements position and signal to noise ration
5	MFP and PTRM	Ranging and depth	Middle and far; the environment parameters accuracy, mismatch
6	Vector method	Bearing ranging and depth	Middle and far; system sensitivity and method based vector sensors
7	Waveguide invariant	Ranging	Near and middle; signal to noise ration and signal bandwidth

The purpose of this paper is to estimate the range of the closest point of approach (CPA) from target to receiving hydrophone (or array), the moving speed of the target, and the angle between the target's moving direction and the array's heading direction. This method needs at least two hydrophones or arrays, for each hydrophone record and then generates the LOFARgrams respectively, then the target motion parameters can be derived from analysis of the LOFARgrams. Through simulation, the correctness of the method will be verified.

We have three hypotheses: first we need at least two hydrophones or arrays for recording the wideband noise of the target; second, the target must go through the CPA point of each hydrophone; third, the target must move with steady speed v.

2 Theory

2.1 Detection Model

The sketch of the moving situation between target and receiving linear array is shown in Fig. 1. The target is moving from the bottom to the upper portion at the left with a consistent speed v, the receiving array is placed along the dashed line, and θ is the angle between the target moving direction and the array bearing. We concern ourselves with the LOFARgrams recorded by the sensors (or sub-array) A and B.

At time t_{0B} and t_{0A}, the target goes through the CPA points of sensors B and A, and the corresponding ranges are D_{CPA_B} and $D_{\mathrm{CPA_A}}$ respectively; at the same time

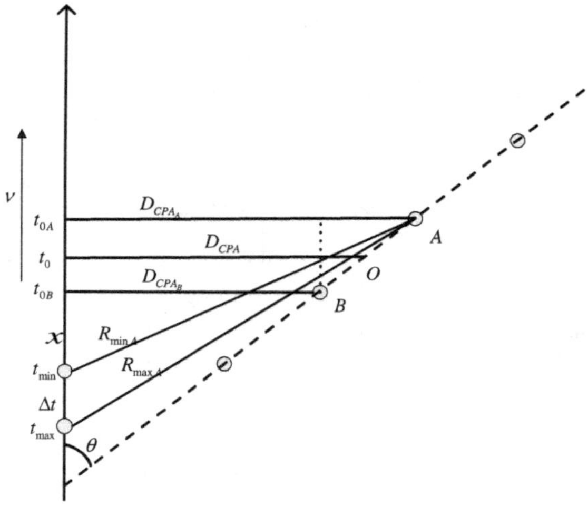

Fig. 1 Sketch of the moving situation between target and receiving linear array

the receiving LOFARgrams will present parabola stripes (as shown in Fig. 2). The center of the sensors A and B (or the center of the whole array) is point O, the range of the CPA is D_{CPA} and the passing time is t_0. Furthermore, when we focus on the time period from $t_{\max}$ to $t_{\min}$, which corresponds to the block area in Fig. 2, the stripes in this block area are nearly straight lines, as the range is decreased monotonously.

2.2 Estimation Method

With the time-frequency of the LOFARgrams shown in Fig. 2, we use the equations

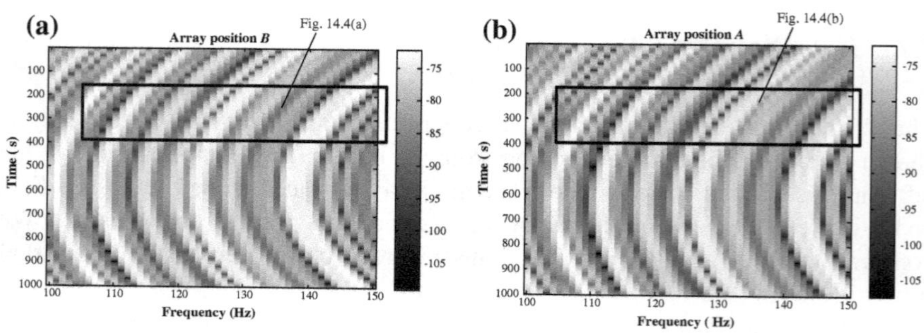

Fig. 2 LOFARgrams recorded by sensors A and B, from the simulation

$$R_{\min A} = \sqrt{D^2_{\mathrm{CPA}_A} + x_A^2} \tag{1}$$

$$R_{\max A} = \sqrt{D^2_{\mathrm{CPA}_A} + (x_A + \Delta t \cdot v)^2} \tag{2}$$

$$R_{\min B} = \sqrt{D^2_{\mathrm{CPA}_B} + x_B^2} \tag{3}$$

$$R_{\max B} = \sqrt{D^2_{\mathrm{CPA}_B} + (x_B + \Delta t \cdot v)^2} \tag{4}$$

where $\Delta t = t_{\max} - t_{\min}, x_A = (t_{0A} - t_{\max}) \cdot v, x_B = (t_{0B} - t_{\max}) \cdot v$

We can get the distance corresponding to the start and end time of this sub area, so based on the WI theory and through extracting the parabolic and straight stripes in Fig. 2, the equations about WI β, speed v and range D_{CPA} can be derived at last for sensors A and B

$$\beta = f(R_{\min A}, R_{\max A}, f_{\min A}, f_{\max A}) = f(v, D_{\mathrm{CPA}_A}) \tag{5}$$

$$\beta = f(R_{\min B}, R_{\max B}, f_{\min B}, f_{\max B}) = f(v, D_{\mathrm{CPA}_B}) \tag{6}$$

On the other hand, through the sketch in Fig. 1, the relationship between the angle θ, and the differences of D_{CPA_B} and D_{CPA_A}, denoted as ΔD, can be determined as follows

$$\sin\theta = \frac{\Delta D}{\Delta d} \tag{7}$$

where Δd is the interval of array elements A and B.

The equation between speed v and angle θ is

$$\cos\theta = \frac{(t_{0A} - t_{0B})v}{\Delta d} \tag{8}$$

With one sensor, we can get one equation, and with two sensors, we can get two equations, such as Eqs. (5) and (6), then, based on the method proposed by Ref. [10], the β, D_{CPA} and v can be obtained. Then the angle θ and the distance ΔD can be derived from Eqs. (7) and (8), and considering relationships:

$$D_{\mathrm{CPA}_B} = D_{\mathrm{CPA}} - \Delta D/2 \tag{9}$$

$$D_{\mathrm{CPA}_A} = D_{\mathrm{CPA}} + \Delta D/2 \tag{10}$$

Then D_{CPA_B} and D_{CPA_A} can be solved respectively.

3 Simulation

The method above will be verified by using a computer simulation. The simulation is conducted with an acoustic toolbox (AT). During the simulation, the target is traveling at a constant speed of 3.23 m/s, the angle between the target moving direction and the array heading direction is $\theta = 37°$, the closest range to the array is $D_{\mathrm{CPA}} = 3500$ m and the interval of A and B is $\Delta d = 200$ m, the center of the array (also the center of A and B) is set to origin, and the target radiation low-frequency bandwidth noise is about 100–150 Hz. The simulation follows the sketch as shown in Fig. 1, and the acoustic field model and basic parameters used in the simulation refer are referred to Fig. 3.

The simulation generates LOFARgrams as shown in Fig. 2; the time and distance corresponding to the peak of the parabola are t_{0A}, t_{0B} and $D_{\mathrm{CPA}_A}, D_{\mathrm{CPA}_B}$, respectively; the parabola stripes can be abstracted using the corresponding method. Figure 4, which is part of Fig. 2, can be used to abstract straight stripes and estimate the WI.

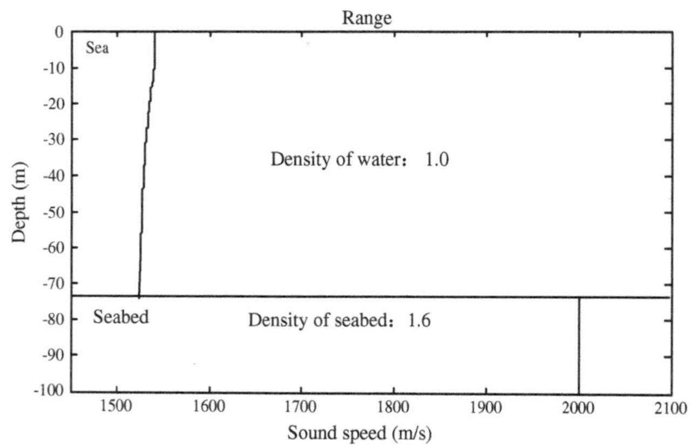

Fig. 3 Simulation acoustic field model

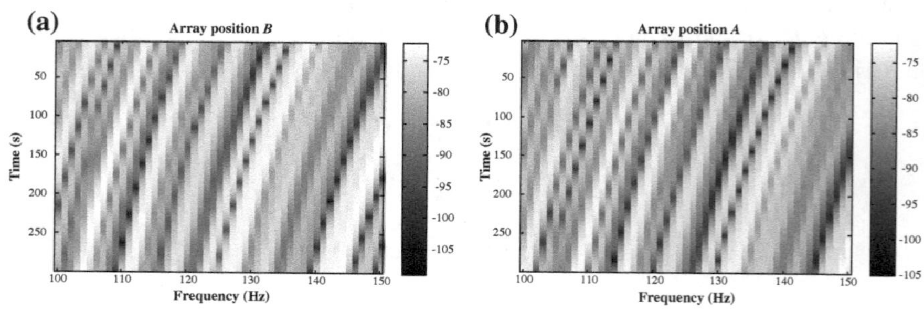

Fig. 4 Part of the LOFARgrams, used for a waveguide invariant abstract

Table 2 List of target parameters, estimated values versus true values

Parameter	Estimated value	True value	Relative error (%)
v	2.92 m/s	3.23 m/s	9.6
θ	43.68°	37°	18
D_{CPA_A}	3648 m	3560 m	2.5
D_{CPA_B}	3510 m	3440 m	2.0
D_{CPA}	3579 m	3500 m	2.3

Such parameters, about time, can be read out directly from Figs. 2 and 4.

From Fig. 2, we obtain the time when the target passes by the CPA point of sensor B: $t_{0B} = 575$ s, and the time when the target passes by the CPA point of sensor A: $t_{0A} = 624$ s. From Fig. 4, we obtain that the start time of the figure is $t_{\min A} = t_{\min B} = 0$ s and the end time of the figure is $t_{\max A} = t_{\max B} = 300$ s. Then, $\Delta t = t_{\max} - t_{\min} = 49$ s and $t_0 \approx 600$ s is also known.

With the values obtained above and the proposed method applied, the unknown parameters can be finally estimated, as shown in Table 2, and as a comparison, the true values are also listed in the table. The WI estimate result is $\beta = 0.885$.

As we can see, by comparing the true values and estimated values in Table 2, the relative error of the estimated distance of the CPA is 2.3 %, while the relative error of the estimated target velocity is 9.6 %, and the relative error of the estimated angle between the target moving direction and the array heading direction is 18 %. The simulation experiment has achieved good results in estimating the target distance, the relative error being less than 5 %, yet the velocity and angle have a higher relative error.

According to the analysis of the simulation results, the target moving angle has the greatest relative estimated error, which is mainly because the angle is calculated from the velocity using Eq. (8), and a small value difference of the velocity or time estimation can lead to a relatively big angle difference.

4 Conclusion

A low-frequency broadband radiated noise for an underwater moving target generates a stable interference structure in the receiving LOFARgrams, which can be described using a WI. In this paper, a method for target parameters (including velocity, heading angle and range) estimation is proposed with the LOFARgrams. This method needs at least two array elements, requires a target with steady speed v and goes through the CPA point of each of the array elements simultaneously.

The method was employed to process the LOFARgrams which were obtained from the AT simulation. The parameters estimation results show the validity of the method. The relative error of the range estimated parameters is less than 5 %.

This method will have a better performance on angle estimation and a better application prospect when combined with other underwater detection methods such as beamforming and vector technology. At the same time, there remain some problems with this method, such as the three hypotheses. The question is whether they can be overcome by method improvement or not. Further research should answer these questions and focus on the case of how to improve the estimation accuracy of the target parameters.

Acknowledgments We thank professor T.C. Yang and W.A. Kuperman for their good advice to this research project during OA 2012. This work was supported by: Open fund of State Key Laboratory of acoustics (SKLA201504); Natural Science Foundation of JiangSu Province of China (No. BK20161195); Central University fund (2013B18514, 2015B25214, 2014B16114). National Natural Science Foundation of China (Grant No. 61273170).

References

1. Li, Q.H.: Principle of Digital Sonar Design. Anhui Education Press, Hefei (2002)
2. Brian, G.F., Kam, W.L.: Passive ranging errors due to multipath distortion of deterministic transient signals with application to the localization of small arms fire. J. Acoust. Soc. Am. **111** (1), 117–128 (2002)
3. Mao, W.N.: An overview of passive localization for under-water acoustics. J. Southeast Univ. **31**(6), 1–4 (2001)
4. Burdic, W.S.: Underwater Acoustic System Analysis. Prentice-Hall, Englewood Cliffs (1991)
5. Xue, S.H., Ye, Q.H., Huang, H.N., et al.: Passive-range estimation using near-field MVDR dual focused beamformers. Appl. Acoust. **24**(3), 177–181 (2005)
6. Shang, E.C., Clay, C.S., Wang, Y.Y.: Passive harmonic source ranging in waveguides by using mode filter. J. Acoust. Soc. Am. **78**(1), 172–175 (1985)
7. Ma, J.G.: Passive Localization Technology of Time Reversal. Ph.D Thesis, Harbin Engineering University (2007)
8. Sun, G.Q., Li, Q.H.: Progress of study on acoustic vector sensor. Acta Acustica **29**(6), 481–490 (2004)
9. Ren, Y., Qi, Y.B.: Waveguide invariant and range estimation based on phase-shift-compensation of underwater acoustic spectrograms. In: Advances in Ocean Acoustics. Proceedings of the 3rd International Conference on Ocean Acoustics, vol. 1495(1), pp. 627–633 (2012)
10. Li, J., Sun, G.Q., et al.: Acoustics vector sensor linear array passive ranging based on waveguide invariant. In: Advances in Ocean Acoustics. Proceedings of the 3rd International Conference on Ocean Acoustics, vol. 1495(1), pp. 576–586 (2012)

Source Localization by Maximizing the Longitudinal Correlation Using Waveguide Invariant Theory

Xian Zhu and Huiliang Ge

Abstract Because of the multipath propagation of an acoustical signal in a shallow water waveguide, the correlation of acoustical signals received by two separated hydrophones declines with longitude distance. In ocean waveguide propagation theory, the waveguide invariant (β) can be used to improve the signal correlation. When using two arrays to replace hydrophones, a source localization method by maximizing the correlation coefficient between one beamformed signal and another beamformed signal modified by a frequency shift is derived in this paper. It is shown by simulation and experiments that the localization error can be less than 10 % when the SNR at the beamformer output is more than 6–9 dB. It is also shown that this method is not suitable to estimate the range of source near the broadside of the line joined by the two array centers.

Keywords Waveguide invariant · Signal correlation · Source localization

1 Introduction

Most research on acoustic localization in the ocean has focused on using coherent signal processing techniques such as matched field processing (MFP) [1–3]. While MFP works well in theory and in numerical simulation, it is not always applicable to real-world situations, because it requires very accurate knowledge of the environment in order to correctly localize the source. Other source localization methods have been developed, requiring much less a priori knowledge about the environment by using the concept of waveguide invariant [4, 5].

In the present study, since the signal correlation can be improved by the waveguide invariant theory [6], we use two sub-arrays to replace two hydrophones. After obtaining the estimation of the bearing and waveguide invariant parameter, an approach for passive range estimation based on maximizing the correlation

X. Zhu (✉) · H. Ge
Hangzhou Applied Acoustics Research Institute, Hangzhou 310012, China
e-mail: sklzhuxian@163.com

L. Zhou et al. (eds.), *Underwater Acoustics and Ocean Dynamics*
DOI 10.1007/978-981-10-2422-1_13

coefficient between one beamformed signal and another beamformed signal modified by frequency shift is derived.

2 Theory

In an ocean waveguide, assume a point source transmits a signal whose spectrum is $S(\omega)$, the received signal spectrum at range r can be expressed as:

$$p(\omega, r) = S(\omega)G(\omega, r) \tag{1}$$

where $G(\omega, r)$ is the channel frequency response function in an ocean waveguide, so the received spectrum at range $r + \Delta r$ is $p(\omega, r + \Delta r)$. When $\Delta r \ll r$, we have $\beta \frac{\Delta r}{r} \ll 1$. If the source spectrum is a slow varying function of the frequency, which means

$$S\left[\omega\left(1 + \beta\frac{\Delta r}{r}\right)\right] \approx S(\omega) \tag{2}$$

According to the waveguide invariant theory, we have [6]

$$p\left[\omega\left(1 + \beta\frac{\Delta r}{r}\right), r + \Delta r\right] \approx \mathrm{e}^{\mathrm{i}\omega\tau} p(\omega, r) \tag{3}$$

where $\tau = \left(\frac{\beta}{c_g^m} + \frac{1}{c_p^m}\right)\Delta r + \frac{\beta}{c_g^m}\frac{(\Delta r)^2}{r}$.

In statistical theory, the signal correlation coefficient of $p(\omega, r)$ and $p(\omega, r + \Delta r)$ can be expressed as:

$$\rho_0 = \max_{\Delta t} \frac{\mathrm{Re}\left\{\int_{\omega_1}^{\omega_2} p(\omega,\ r)p^*(\omega,\ r + \Delta r)\mathrm{e}^{-\mathrm{i}\omega\Delta t}\mathrm{d}\omega\right\}}{\sqrt{\int_{\omega_1}^{\omega_2} |p(\omega,\ r)|^2 \mathrm{d}\omega \int_{\omega_1}^{\omega_2} |p(\omega,\ r + \Delta r)|^2 \mathrm{d}\omega}} \tag{4}$$

where $\mathrm{Re}\{\cdot\}$ represents the real part. Because of the interference of the normal mode, the correlation coefficient is obviously less than 1 when Δr is increasing to a certain value. To improve the signal correlation, a new correlation coefficient is defined as:

$$\rho_1 = \max_{\Delta t} \frac{\mathrm{Re}\left\{\int_{\omega_1}^{\omega_2} p(\omega, r)p^*\left[\omega\left(1 + \beta\frac{\Delta r}{r}\right),\ r + \Delta r\right]\mathrm{e}^{-\mathrm{i}\omega\Delta t}\mathrm{d}\omega\right\}}{\sqrt{\int_{\omega_1}^{\omega_2} |p(\omega, r)|^2 \mathrm{d}\omega \int_{\omega_1}^{\omega_2} \left|p\left[\omega\left(1 + \beta\frac{\Delta r}{r}\right),\ r + \Delta r\right]\right|^2 \mathrm{d}\omega}} \tag{5}$$

where $p[\omega(1 + \beta\Delta r/r), r + \Delta r]$ is a frequency shift compensation from $p(\omega, r + \Delta r)$. It has been proven that $\rho_1 \approx 1$ if Eq. (3) is satisfied. Therefore, the signal correlation can be improved significantly after the frequency shift compensation with waveguide invariant (β), range (r), and the range variation (Δr).

A horizontal line array (HLA) which has N elements spaced d, receives a signal from bearing θ. If we divide the horizontal array into two equal sub-arrays, we can get two output signals $p_b(\omega, r)$ and $p_b(\omega, r + Nd\cos\theta/2)$ from the sub-array beamforming. When we use two sub-arrays to replace two hydrophones, we can get a correlation coefficient of two beam output signals as in Eq. (5).

$$\rho(\beta, r, \theta) = \max_{\Delta t} \frac{\mathrm{Re}\left\{\int_{\omega_1}^{\omega_2} p_b(\omega, r) p_b^*\left[\omega\left(1 + \beta\frac{Nd\cos\theta}{2r}\right),\ r + \frac{Nd\cos\theta}{2}\right] \mathrm{e}^{-\mathrm{i}\omega\Delta t} \mathrm{d}\omega\right\}}{\sqrt{\int_{\omega_1}^{\omega_2} |p_b(\omega, r)|^2 \mathrm{d}\omega \int_{\omega_1}^{\omega_2} \left|p_b\left[\omega\left(1 + \beta\frac{Nd\cos\theta}{2r}\right),\ r + \frac{Nd\cos\theta}{2}\right]\right|^2 \mathrm{d}\omega}} \tag{6}$$

where the bearing θ can be estimated by array beamforming, the waveguide invariant can be estimated by the mathematical definition of a waveguide invariant [7, 8] based on the normal mode propagation model or extracted from the experimental data. So the target range can be estimated using,

$$\hat{r} = \max_{r} \rho(\hat{\beta}, r, \hat{\theta}) \tag{7}$$

3 Simulations

Source range estimation by the waveguide invariant methods is illustrated by a simulation example involving typical ocean waveguide environments and array configurations. The example employs a receive horizontal line array in a Pekeris waveguide where the sea depth is 100 m, the sound speed in the water changes linearly from 1500 m/s at sea surface to 1490 m/s at sea bottom, and the sound speed, density, and attenuation of the sea sediment are 1637 m/s, 1.793 g/cm^3 and 0.2 dB/λ, respectively. The source radiates 650–750 Hz noise at a depth of 30 m. The receive array has 360

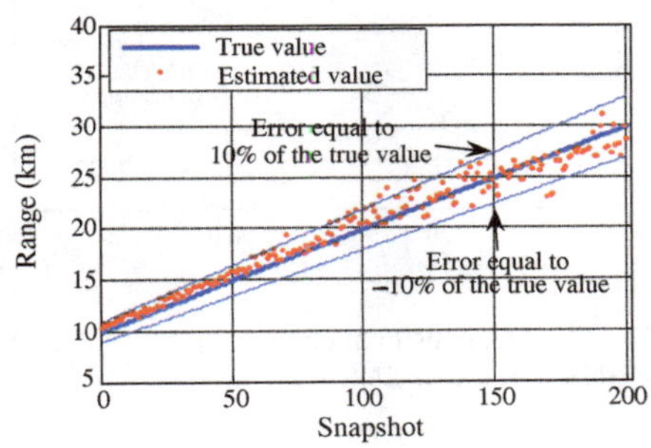

Fig. 1 Range estimation results for source at 110°

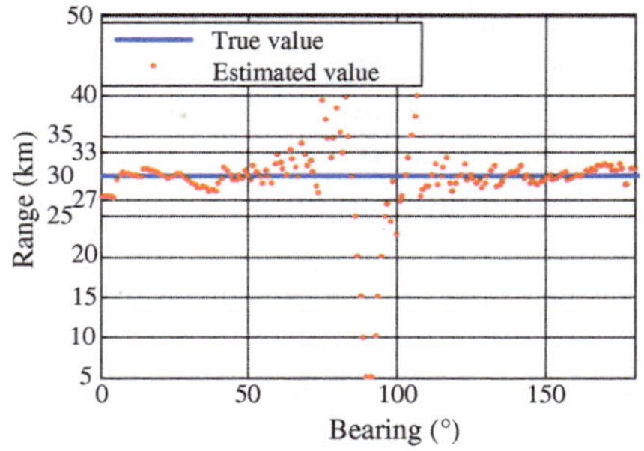

Fig. 2 Range estimation results for source at different bearings

elements spaced 1 m at a depth of 25 m. The waveguide invariant is 0.99 which can be calculated using a KRAKEN normal model program [9].

In the case of a large SNR, first, assuming one source moves from 10 to 30 km, the range estimation results are shown in Fig. 1 when the source bearing is 110°. Next, assuming one source moves from 0° to 180°, the range estimation results are shown in Fig. 2 when the source range is 30 km. The results of these two figures

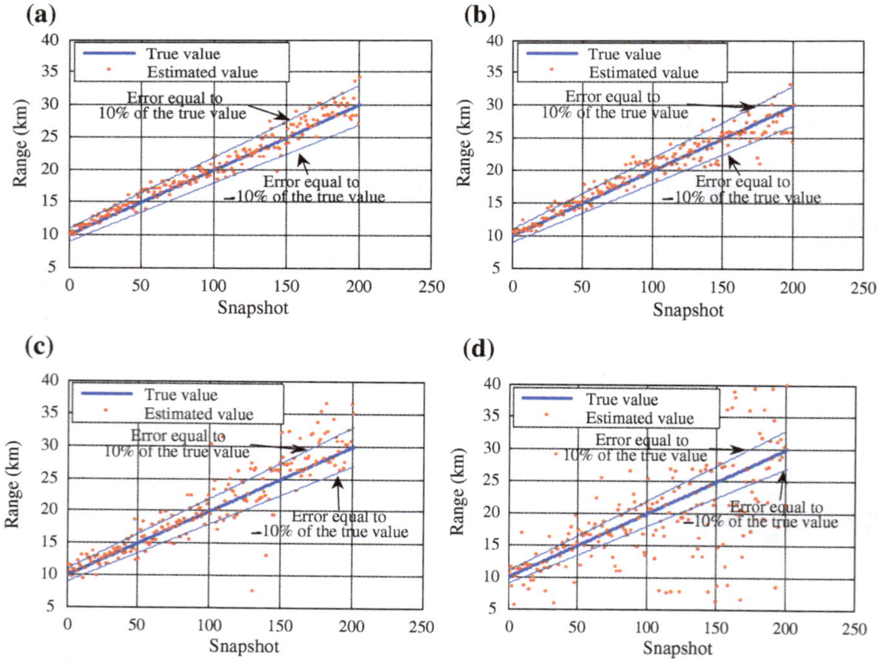

Fig. 3 Range estimation results for different SNR cases. **a** SNR = 12 dB; **b** SNR = 9 dB; **c** SNR = 6 dB; **d** SNR = 12 dB

show that, at a fixed bearing, the range estimation error is less than 10 % of the source whose range is within 30 km, and the estimated error will increase when the source range increases; at a fixed range, the estimated error will increase when source moves from the end fire (0° or 180°) to the broadside (90°). Especially, when the source bearings are in the range of 80–100°, the estimated error will increase rapidly.

The range estimated results in different SNR cases are shown in Fig. 3 when the source's bearing is 135° in a white noise background. It is shown that the estimated error increases with the SNR becoming smaller. If the estimated error is less than 10 % of the source whose range is within 30 km, the SNR should be more than 6–9 dB.

4 Experimental Data Analysis

The performance of the waveguide invariant method for range estimation with experimental data is given. The depth of the experimental sea is 80 m, and the sound speed profile is plotted in Fig. 4. A horizontal receiver array was towed along linear tracks with a velocity of 5 m/s at a fixed depth of 25 m. A static source radiated 650–750 Hz noise with a depth of 30 m. The waveguide invariant is about 0.99 which was calculated using a KRAKEN normal model program [9]. The source bearing varied from 132° to 143°, and the SNR of the beam output is about 6 dB. The range estimation results using a waveguide invariant and the range measurement result from the GPS are given in Fig. 5. It can be found that the estimated errors of most of the range points are within 10 % of the source range which varied from 14 to 18 km.

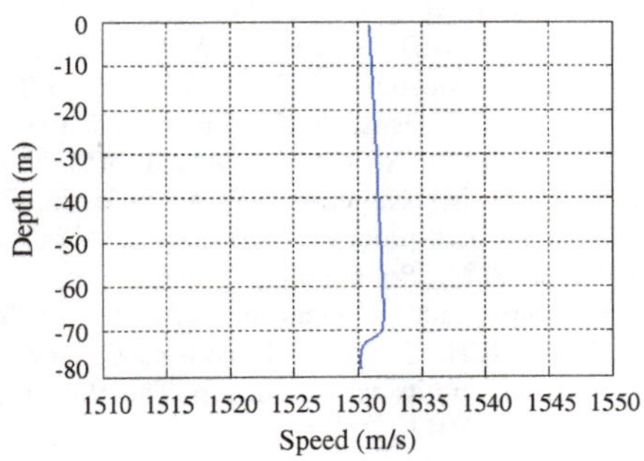

Fig. 4 Sound–speed profile measured

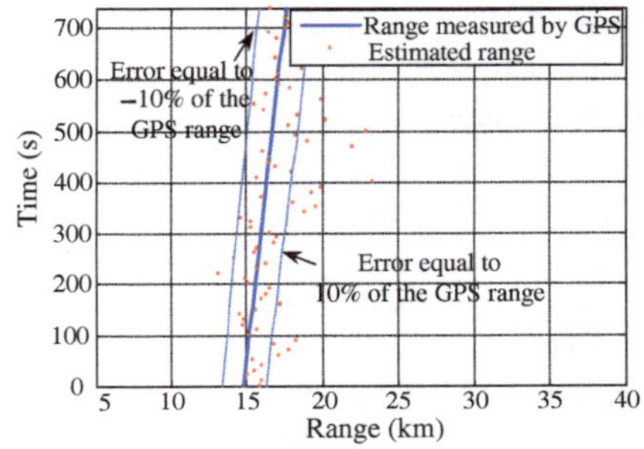

Fig. 5 Range estimation results for sea experimental data

5 Conclusion

An approach for passive range estimation based on improvements in the correlation for two horizontal sub-array signal beam outputs is derived after obtaining the estimation of the bearing and waveguide invariant parameter. It is shown that this range estimation approach only requires a waveguide invariant value and is robust to environmental parameters. By simulation analysis and the validation of the experimental array data at sea with a 360-m horizontal line array in shallow water, we found that the range estimation error using the derived approach is less than 10 % of the source whose range is within 30 km when the SNR at the beamformer output is more than 6–9 dB. However, there is an estimated blind zone at angles near the broadside that are about 20° for a horizontal array with a 360-m aperture.

References

1. Bucker, H.P.: Use of calculated sound fields and matched-detection to locate sound source in shallow water. J. Acoust. Soc. Am. **59**(2), 368–373 (1976)
2. Baggeroer, A.B., Kuperman, W.A., Mikhalevsky, P.N.: An overview of matched field methods in ocean acoustics. IEEE J. Ocean Eng. **18**(4), 401–424 (1993)
3. Soares, C., Jesus, S.M.: Broadband matched-field processing: Coherent and incoherent approaches. J. Acoust. Soc. Am. **113**(5), 2587–2598 (2003)
4. Thode, A.M., Kuperman, W.A., D'Spain, G.L., Hodgkiss, W.S.: Localization using Bartlett matched field processor sidelobes. J. Acoust. Soc. Am. **107**(1), 278–286 (2000)
5. Thode, A.M.: Source ranging with minimal environmental information using a virtual receiver and waveguide invariant theory. J. Acoust. Soc. Am. **108**(4), 1582–1594 (2000)
6. Zhang, R.H., Li, F.H.: Reconstruction of low frequency pulse waveform with waveguide invariant and its application. In: The 19th International Congress on Acoustics, Madrid, Spain, pp. 7039–7043 (2007)
7. Chuprov, S.D.: Interference structure of a sound field in a layered ocean. In: Ocean Acoustics, Nanka Moscow, pp. 71–91 (1982)
8. Brekhovskikh, L.M., Lysanov, Y.P.: Fundamentals of ocean acoustics, 2nd edn. Springer, New York (1991)
9. Porter, M.B.: The Kraken normal model program. Naval Research Lab, Washington, DC (1992)

Selective Detection and Localization by Decomposition of a Subrank Time Reversal Operator

Chunxiao Li, Mingfei Guo and Huancai Lu

Abstract In this paper, a laboratory experiment for the detection and localization of targets using a subrank beam-based decomposition of the time reversal operator (DORT) method is presented. A rigid 32-element source-receiver array, connected to a programmable electronic system, was deployed in a 1.44-m-deep water tank. Eight snapshots, much less than the number of the array, were collected to construct a covariance matrix for replacing the time reversal operator. The experimental setup consisted of placing a non-pointlike scatterer and a pointlike scatterer at the same range but at different depths. It was shown that focusing is only achieved on each target at a frequency with a relatively large singular value. Due to the fact that the largest target is apparently stronger than the smallest one, the coupling between targets has a minor effect on the quality of the focusing or on the position of the largest target. The frequencies that can achieve good focusing occupancy are three-fourths of the whole frequency band. However, the ratio is significantly decreased when the second singular vector is numerically backpropagated.

Keywords Decomposition of the time reversal operator · Target detection · Target localization · Subrank time reversal operator · Underwater acoustic signal processing

---

C. Li (✉) · M. Guo · H. Lu
MOE Key Laboratory of Mechanical Manufacture and Automation,
College of Mechanical Engineering, Zhejiang University of Technology,
Hangzhou 310014, China
e-mail: chunxiaoli@zju.edu.cn

C. Li · H. Lu
Zhejiang Key Laboratory of Signal Processing, Zhejiang University of Technology,
Hangzhou 310014, China

L. Zhou et al. (eds.), *Underwater Acoustics and Ocean Dynamics*
DOI 10.1007/978-981-10-2422-1_14

1 Introduction

The decomposition of the time reversal operator (DORT under its French language acronym) has been intensively investigated since it was proposed in 1995 [1]. In a practical application, the measurement of the array response matrix is an important step for the subsequent signal processing. The original DORT method is based on an element-to-element data matrix [2, 3]. This approach is to measure the response matrix by sequentially transmitting signals from individual source elements and recording the backscattered echo on each receiver element. In order to get a sufficient signal-to-noise ratio, this method is extended to the beamspace basis for the use of transmitted and received beams [4, 5].

However, both forms of DORT mentioned above are implemented to construct a full-rank matrix. The number of snapshots required is equal to that of the source-receiver array (SRA). Considering that the ocean embodies time-varying processes, the performance of the DORT method could possibly be degraded due to the acquirement time of the response signals exceeding the coherent of the medium. A more practical subrank DORT method has been proposed and demonstrated in simulation during recent years [6]. Compared to the full-rank DORT, a few snapshots are needed to construct the time reversal operator (TRO). This paper focuses on the application of the subrank DORT method to separate and focus numerically on two targets in very shallow water in the presence of reverberation. This paper differs from Ref. [6] in a number of aspects: (1) reverberation is considered in this paper, while the simulation is free of bottom reverberation with limited experimental data using an echo repeater to simulate the target response in Ref. [6]; (2) except for a single target, the separation and selective focusing on two targets are investigated using laboratory measurement in a water waveguide.

2 Theory

The principle of the DORT is described in several papers [1–3]. It requires the measurement of the array response matrix $k_{jl}(\omega)$, i.e., the backscattered field on channel j of the SRA due to an impulsive signal broadcast from channel l. Obviously, the array response matrix is made up of the $N \times N$ interelement impulse responses of the array ($N = 32$ in the experiment). The TRO is defined as $K^{\mathrm{H}}(\omega)K(\omega)$ from the transmit array point of view and $K(\omega)K^{\mathrm{H}}(\omega)$ from the receive point of view, where the superscript H denotes a complex transpose.

The element-to-element measurement of the transfer matrix is difficult to obtain due to significant background noise. Therefore, a beam-based DORT is proposed in Ref. [5], which uses N orthogonal beams of the Hadamard basis. As E represents a matrix composed of a complete set of orthogonal array weighting vectors, the beam response matrix $\widetilde{K}_{jl}(\omega)$ is the backscattered field received on beam *ej* from the impulsive signal transmitted on beam *el*. The relationship between the beam response matrix and the element transfer matrix is $\widetilde{K}_{jl}(\omega) = E^{\mathrm{T}}K(\omega)E$, where $E^{\mathrm{T}}E = EE^{\mathrm{T}} = I$ and the superscript T denotes transpose. Both of the above methods will obtain a full-rank TRO.

In fact, the final plan to form a TRO is to estimate the Green's function of the target. A complete, orthogonal basis is not necessarily required. The covariance matrix used in matched filed processing is expressed as

$$C(\omega) = \sum_{m=1}^{M} r_m(\omega) r_m^{\mathrm{H}}(\omega) \quad (1)$$

where $r_m(\omega)$ is the *m*th snapshot of the complex pressure measured along the SRA at an angular frequency ω. The covariance matrix is equal to the TRO from the receive point of view if *M* is equal to the number of the array elements and the snapshots are recorded by sequentially transmitting signals from the individual source elements [6].

3 Experimental Setup and Signal Processing

The water tank was 14 m in length, 1.2 m in width and 1.44 m in height (Fig. 1). Three sides were covered with anechoic tile and the other side was a steel sheet located 12 m away from the SRA. The bottom was covered with sand 0.22 m thick. The environmental parameters are as follows: in the water column, density $\rho_1 = 1000\ \mathrm{kg/m^3}$, and wave speed $c_1 = 1490$ m/s; at the bottom layer, sand density $\rho_2 = 1800\ \mathrm{kg/m^3}$, sand speed $c_2 = 1650$ m/s, and attenuation $\alpha_2 = 0.67$ dB per wavelength. The experiment was performed using a vertical SRA of 32 elements

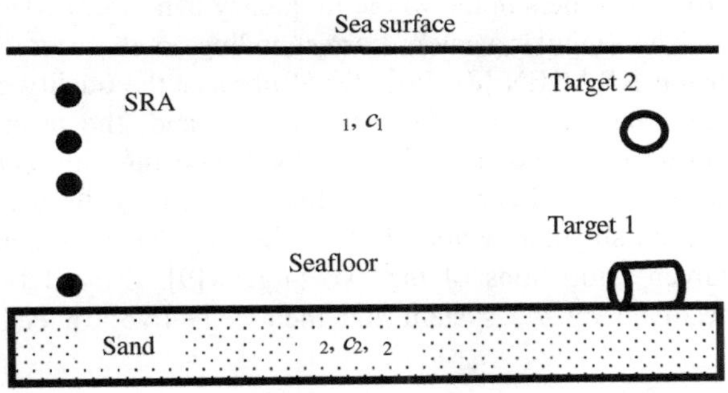

Fig. 1 Experimental setup and model for the acoustic environment

equally spaced at 0.04 m apart. Each of the 32 elements was individually controlled and amplified during transmission and reception. There are two targets: the first one is an air-filled, steel cylindrical shell of 0.21 m in diameter and 0.51 m in length; the other is an air-filled steel spherical shell of 0.10 m in diameter. A broadside transmission is carried out from the SRA for the LFM signal with a center frequency of 12 and a 4-kHz frequency bandwidth. The two targets mentioned above are located at the same range (about 9 m ($\sim 72\lambda$)), but different depths of 0.5 m ($\sim 4\lambda$) for the spherical shell and along the bottom for the cylindrical shell. There were only eight snapshots used for constructing the TRO. The separation and localization of the two targets using a subrank DORT were investigated.

After an ensemble of data snapshots $r_m(t)$ is collected, the received data can be arranged as a matrix $K(t) = [r_1(t), r_2(t), \ldots, r_M(t)]$, and the TRO is defined as $H(t) = K(t)K(t)^{\mathrm{H}}$ for a full-rank DORT or covariance matrix (Eq. 1) for subrank DORT. The TRO $H(t)$ is a $N \times N$ Hermitian matrix, and the recording time t corresponds to the round trip propagation from the SRA to the detection distance $d = tc/2$, where c is the speed of the sound in the waveguide. In order to get a sufficient signal-to-reverberation ratio, the TRO in the frequency domain is built from short time windows. That is, the matrix $H(\omega)$ is the Fourier transform $H(t)$ between time t and $t + \Delta t$. For a practical application of the DORT, the choice of Δt depends on the acoustic propagation characteristics of the waveguide (e.g., the disperse effect) and the range from the SRA to the target, which should include the complete response of the target and minimize the effects of reverberation.

4 Experimental Results

According to Ref. [7], the depth resolution at 3 dB in the waveguide described in Fig. 1 for the LFM signal in the 10–14 kHz frequency band is about 0.12–0.16 m, so that the two targets are well resolved by the SRA. It appears that the first two singular values have apparent peaks emerging at the distance of 8.6 m as shown in Fig. 2. For this distance, the two dominant singular values are clearly separated from the others in the whole frequency band from 10 to 14 kHz, as shown in Fig. 3.

The singular vectors corresponding to these are respectively backpropagated using KRAKEN [8]. In order to observe the quality of the focusing both in depth and range for the whole frequency band, the peak positions of the ambiguity surfaces are shown in Fig. 4. Good focusing only occurs in part of the frequency band with a larger singular value (Fig. 3) for the backpropagation of the first and second singular vectors. In fact, the singular vectors are linear combinations of the Green's functions of the two targets [9]. Only if two pointlike targets are well resolved by the system and their "effective" reflectivities are different, the first

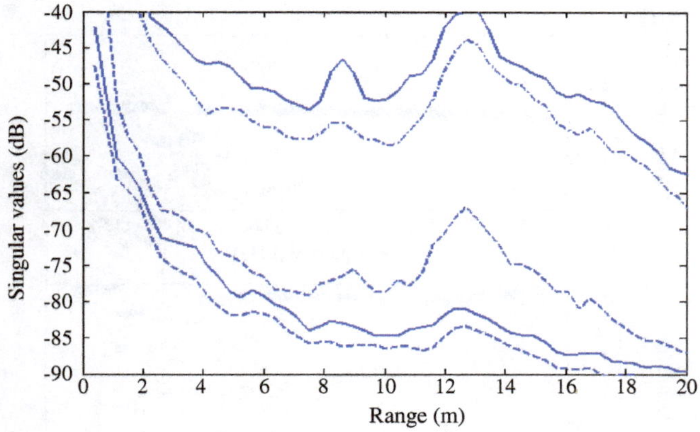

Fig. 2 Two targets at a range of 9 m: the first five singular values as a function of distance from 0 to 20 m

Fig. 3 The singular values as a function of the frequency from 10 to 14 kHz at the distance of 9 m for two targets at different depths. **a** All singular values; **b** the first singular value; **c** the second singular value

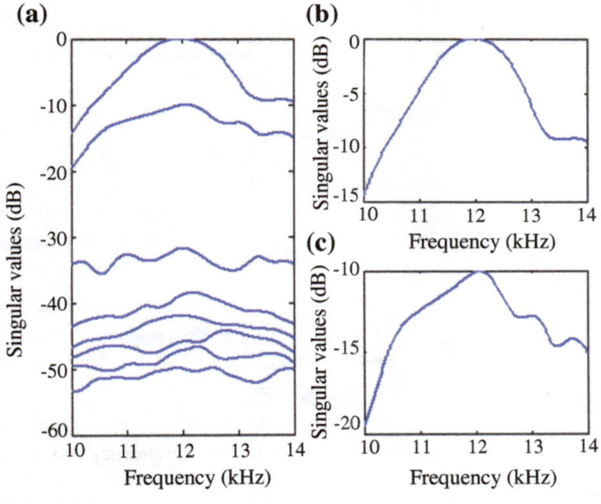

singular vector is approximately associated to the stronger target and the second singular vector approximately to the weaker one. Since the largest target is non-pointlike, a coupling between the targets cannot be ignored. However, due to the largest target apparently being stronger than the smallest one, the focusing quality of the first singular vector on the largest target is superior to that of the smallest one. As seen in Fig. 4, the frequency band that can achieve good focusing on the position of the largest target is 10.3–12.9 kHz, while it is 11–12 kHz for the smallest target.

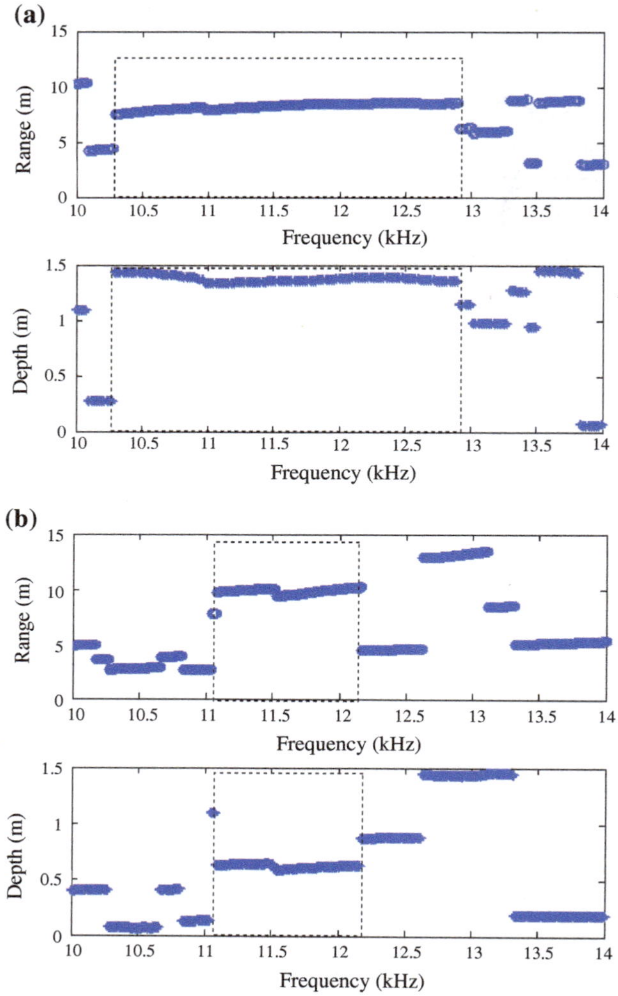

Fig. 4 The peak positions of the ambiguity surfaces for backpropagating **a** the first and **b** the second singular vectors in the whole frequency band for the case of two targets. Good focusing for each target is shown in the *dotted frame*

5 Conclusion

In this paper, the feasibility of target detection and localization using the subrank beam-based DORT was demonstrated by experiments in a laboratory waveguide. Compared to full-rank DORT, the subrank beam-based DORT requires a shorter time to develop the TRO. The detection and localization of two well-resolved scatterers can be obtained with a few number of snapshots from a discrete set of transmission beams to ensonify a region of interest.

Due to the fact that the largest target is a non-pointlike scatterer, the dominant singular vectors corresponding to the singular values and the targets are no larger than a one-to-one corresponding relationship. Generally speaking, good focusing occurs in the frequency band with the larger singular value. The impact of the smallest target on the largest one is insignificant. Good focusing on the largest target is achieved in three-quarters of the whole frequency band. However, the backpropagation of the second singular vector can only be numerically focused near the position of the smallest target in a quarter of the whole frequency band.

Finally, we should point out that much fewer snapshots were needed to construct the TRO in these experiments. It has a potential advantage over the full-rank DORT method for practical applications in a real shallow water ocean environment. For future studies, how to choose the number of snapshots and take advantage of the useful frequency to focus on a target should be investigated.

Acknowledgments The authors gratefully acknowledge the support for the waveguide experiment from Hangzhou Applied Acoustics Research Institute. This work was supported by the Natural Science Foundation of China (grant No. 61101231), Science and Technology Department of Zhejiang Province Public Welfare Projects (grant No. 2013C31003) and Zhejiang Provincial Natural Science Foundation of China (grant No. Y1100629).

References

1. Prada, C., Thomas, J.L., Fink, M.: The iterative time reversal process: analysis of the convergence. J. Acoust. Soc. Am. **97**(1), 62–71 (1995)
2. Li, C.X., Li, J.L., Gong, X.Y., Guo, M.F.: Focusing with the decomposition of the time reversal operator method in noisy environments. J. Zhejiang Univ. Sci. A **10**, 997–1008 (2009)
3. Li, J.L., Zhao, H.F., Fang, W.: Experimental investigation of selective localization by decomposition of the time reversal operator and subspace-based technique. IET Radar Sonar Navig. **2**(6), 426–434 (2008)
4. Li, C.X., Xu, W., Li, J.L., Gong, X.Y.: Time-reversal detection of multidimensional signals in underwater acoustics. IEEE J. Ocenic. Eng. **36**(1), 60–70 (2011)
5. Lingevitch, J.F., Song, H.C., Kuperman, W.A.: Time reversed reverberation focusing in a waveguide. J. Acoust. Soc. Am. **111**(6), 2609–2614 (2002)
6. Edelmann, G.F., Lingevitch, J.F., Gaumond, C.F., Fromm, D.M., Calvo, D.C.: Comparison of a subrank to a full-rank time-reversal operator in a dynamic ocean. J. Acoust. Soc. Am. **122**(5), 2706–2714 (2007)
7. Kim, S., Edelmann, G.F., Kuperman, W.A., Hodgkiss, W.S., Song, H.C.: Spatial resolution of time-reversal arrays in shallow water. J. Acoust. Soc. Am. **110**(2), 820–829 (2001)
8. Porter, M.B.: The KRAKEN normal mode program. SACLANTCEN Memorandum SM-245 (1991)
9. Prada, C., Manneville, S., Spoliansky, D., Fink, M.: Decomposition of the time reversal operator: detection and selective focusing on two scatterers. J. Acoust. Soc. Am. **99**(4), 2067–2076 (1996)

Sounds of Undersea Gas Leaks

Alexey O. Maksimov, Boris A. Burov, Alexander S. Salomatin and Denis V. Chernykh

Abstract A passive acoustic method of detecting potentially environmentally disastrous gas leaks from pipelines, and methane naturally leaking from the seabed has been studied. Gas escape involves the formation and release of bubbles of different sizes. Each bubble of a given size emits a sound at a specific frequency. The traditional approach based on the determination of the bubble radius from the frequency of its signature passive acoustic emission by use of so-called Minnaert formula has a restricted area of applicability near the seabed. The point is that the inertial mass of the birthing bubble is markedly different from the one of a free bubble. It was shown that the bi-spherical coordinates provide separation of variables and are most suitable for analysis of volume oscillations of constrained bubble. The theoretical model for the bubble volume oscillations near the seabed has been proposed and an analytical solution has been derived.

Keywords Gas leaks · Bubbles · Volume oscillations · Acoustic emission

1 Introduction

Undersea gas leaks from pipelines and natural hydrocarbon seeps represent a significant hazard to environment. Gas escape involves the formation and release of bubbles of different sizes that can be detected efficiently using passive acoustic techniques. When a leak occurs, each bubble produces a note [1]. The traditional approach is based on the determination of the bubble radius from the frequency of its signature passive acoustic emission by use of so-called Minnaert formula [2]. Therefore, by analyzing the sound you can "theoretically" know how many bubbles are produced and what size they are. The approach is based on a set of equations

A.O. Maksimov (✉) · B.A. Burov · A.S. Salomatin · D.V. Chernykh
Pacific Oceanological Institute, Far Eastern Branch of Russian Academy of Sciences, Vladivostok 690041, Russia
e-mail: maksimov@poi.dvo.ru

L. Zhou et al. (eds.), *Underwater Acoustics and Ocean Dynamics*
DOI 10.1007/978-981-10-2422-1_15

that can be used to write a computer program for solution of the inverse problem that is to interpret those sounds in terms of the bubble size distribution.

This study focuses primarily on the proper description of the sound field generated by individual bubbles. The point is that the inertial mass of the birthing bubble near the seabed is markedly different from the one of a free bubble. We continue the study of this problem, which began in Ref. [3], where an analytical solution has been derived for a perfectly rigid bottom. Here, we would like to account for the elasticity of sediments. The necessity to improve the model emerging due to a series of field experiments in the coastal area of the Sea of Japan. Observations of seabed bubbles at artificial marine seeps by passive acoustic and optical approaches have been compared. Analysis of the recorded signals and attempts to compare them with the predictions of the theory did not lead to an unambiguous result for bubbling through the sandy layer. A preliminary analysis shows that the model of the rigid bottom is not fully suitable for describing the acoustic emission of bubbles birthing in the sandy sediments.

2 Model

Consider a bubble suspended in an ideal fluid (medium 1) with density ρ_1. The bubble is at distance h from an elastic solid (medium 2) with density ρ_2, bulk modulus K and shear modulus μ. The basic bubble equilibrium shape is the sphere of radius R_0. It will be assumed that the size of the bubble is smaller than the emitted acoustic wave length, leading to the velocity being expressed as the gradient of a potential $\varphi(V = \nabla\varphi, \nabla^2\varphi = 0)$. The Bernoulli equation $P(\boldsymbol{r}, t) + \rho_1 [\dot{\varphi}(\boldsymbol{r}, t) - \dot{\varphi}_\infty(t)] = P_\infty$ governs the variation of the pressure in the liquid $P(\boldsymbol{r}, t)$ (P_∞ is the equilibrium pressure).

The dynamic boundary condition is that the pressure on two sides of the bubble wall S differs only because of surface tension $P_l = P_g - \sigma(\nabla \cdot \boldsymbol{n})$, where $\boldsymbol{n}$ is the unit normal, σ is the coefficient of surface tension of the gas/liquid interface; $P_l = P(\boldsymbol{r} \in S, t)$ and $P_g = P_0(V_0/V)^\gamma$ denote the pressure in the liquid and in the bubble respectively. We adopt a polytropic law for the gas in the bubble and V, V_0 are the instantaneous and equilibrium bubble volume, γ is the polytropic exponent, P_0 is the equilibrium pressure in the bubble. The kinematic boundary condition at the bubble wall takes the form of $\dot{\zeta} = (\boldsymbol{v} \cdot \boldsymbol{n})_{\mathbf{r} \in S}$, where ζ is the normal displacement of the bubble wall S.

At this point it is convenient to introduce bi-spherical coordinates admitting separation of variables in the Laplace equation. This coordinate system has been successfully used in Ref. [4] to describe bubble dynamics near a free surface. The toroidal coordinate system, which is very similar to the bi-spherical one, has been successfully applied to analyze the dynamics of constrained bubbles [5, 6]. The bi-spherical coordinates (ξ, ϑ, α) are related to Cartesian ones (x, y, z) by the relations:

$$x = a\frac{\sin\vartheta\cos\alpha}{\cosh\xi - \cos\vartheta}, \quad y = a\frac{\sin\vartheta\sin\alpha}{\cosh\xi - \cos\vartheta}, \quad z = a\frac{\sinh\xi}{\cosh\xi - \cos\vartheta},$$
$$a = R_0\sinh\xi_0, \quad \xi_0 = \ln\left[h/R_0 + \sqrt{h^2/R_0^2 - 1}\right] \tag{1}$$

where ξ goes from $-\infty$ to ∞, and ϑ and α go from 0 to π and from 0 to 2π, correspondingly. The surface $\xi = \xi_0$ is the bubble, while the undisturbed interface corresponds to $\xi = 0$. For $\xi = 0, \cos\vartheta \to 1$ corresponds to $r \to \infty$.

Thus we should find the solution of the Laplace equation, which in bi-spherical coordinates has the following form:

$$\nabla^2\varphi = \frac{1}{h_\xi^3}\left[\frac{1}{\sin\vartheta}\frac{\partial}{\partial\vartheta}\left(h_\vartheta\sin\vartheta\frac{\partial\varphi}{\partial\vartheta}\right) + \frac{\partial}{\partial\xi}\left(h_\xi\frac{\partial\varphi}{\partial\xi}\right) + \frac{h_\xi}{\sin^2\vartheta}\frac{\partial^2\varphi}{\partial\alpha^2}\right] = 0 \tag{2}$$

where $h_\xi = h_\vartheta = a/(\cosh\xi - \cos\vartheta)$ and $h_\alpha = a\sin\vartheta/(\cosh\xi - \cos\vartheta)$ are the Lamé coefficients. To separate the variables in the Laplace equation, we replace $\varphi(\xi, \vartheta, \alpha, t)$ by $\tilde{\varphi}(\xi, \vartheta, \alpha, t)$: $\varphi = \varphi_\infty + \sqrt{\cosh\xi - \cos\vartheta}\,\tilde{\varphi}$. Then, in view of the assumed axial symmetry of the system, Eq. (2) acquires the following form:

$$\frac{1}{\sin\vartheta}\frac{\partial}{\partial\vartheta}\left(\sin\vartheta\frac{\partial\tilde{\varphi}}{\partial\vartheta}\right) = -\frac{\partial^2\tilde{\varphi}}{\partial\xi^2} + \frac{\tilde{\varphi}}{4} \tag{3}$$

The Legendre functions $P_n(\cos\vartheta)$ are the eigenfunctions of the linear differential operator of the left hand side of Eq. (3) with eigenvalues $-n(n+1)$ (n is an integer). The solution of Eq. (3) can be found in the form as follows:

$$\tilde{\varphi}(\xi, \vartheta, t) = \sum_{n=0}^{\infty}\left[a_n e^{-(n+1/2)\xi} + b_n e^{(n+1/2)\xi}\right]P_n(\cos\vartheta) \tag{4}$$

where a_n and b_n should be determined from the boundary conditions.

3 Rigid Bottom Solution

The rigid bottom condition is easily satisfied because it avoids solutions in an elastic half-space. The pressure within the bubble is practically constant, when its size is smaller than the wavelength (homobaric bubble) and hence its surface is equipotential, if neglecting surface tension. We shall analyze linear volume oscillations of the millimeter-sized bubbles and therefore can use this approximation. As the potential is defined within an arbitrary function of time, we choose normalization of this variable to be zero at the bubble wall. The solution, satisfying the boundary conditions at the bubble wall ($\xi = \xi_0$): $\varphi(\xi,\vartheta,t)|_{\xi=\xi_0} = 0$ and at the rigid bottom ($\xi = 0$): $\left.\frac{\partial\varphi(\xi,\vartheta,t)}{\partial\xi}\right|_{\xi=0} = 0$, has the form in Ref. [3]:

$$\varphi(\xi,\vartheta,t) = \varphi_\infty(t)\Big[1 - \sqrt{\cosh\xi - \cos\vartheta}\sqrt{2}\sum_{n=0}^{\infty}\frac{e^{-(n+1/2)\xi_0}}{\cosh[(n+1/2)\xi_0]} \times\cosh[(n+1/2)\xi]P_n(\cos\vartheta)\Big] \quad (5)$$

By calculating the variation of the bubble volume ΔV, and substituting this value in the dynamic boundary condition at the bubble wall, we obtain an analogue of the Rayleigh equation:

$$\begin{aligned} &\ddot{\varphi}_\infty + \Omega_*^2\varphi_\infty = 0 \\ &\frac{\Omega_*^2}{\Omega_0^2} = \left\{1 - 2\sqrt{\kappa^2-1}\sum_{n=0}^{\infty}\frac{\left[\kappa+\sqrt{\kappa^2-1}\right]^{-(2n+1)}}{\left[\kappa+\sqrt{\kappa^2-1}\right]^{(2n+1)}+1}\right\} \\ &\Omega_0^2 = \frac{3\gamma P_0}{\rho_1 R_0^2} \end{aligned} \quad (6)$$

where Ω_* and Ω_0 are the natural frequencies of the bubble located at the distance h above the rigid bottom and a free bubble, respectively, and $\kappa = h/R_0$. It should be noted that the presence of a rigid boundary decreases the natural frequency. A graph of the ratio Ω_*/Ω_0 as function of h/R_0 is shown in Fig. 1a.

For rising bubbles, the distance h can be evaluated as $h = R_0 + ut$, where the rising velocity of millimeter-sized bubbles u equals to 14–25 cm/s. For this reason,

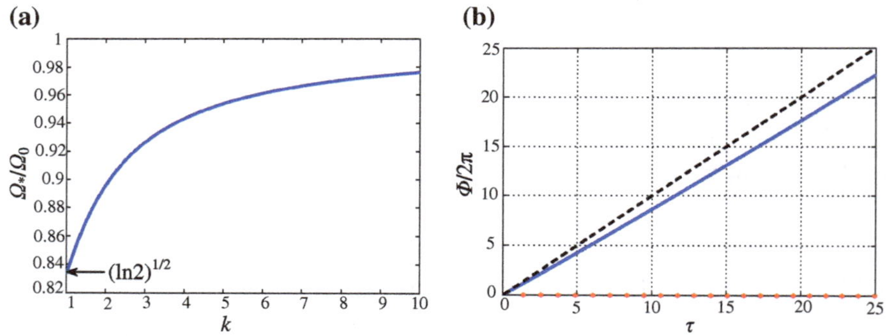

Fig. 1 **a** The variation of the bubble natural frequency with the distance to the rigid bottom. The *solid line* represents the ratio of the oscillation frequency of a bubble near a rigid boundary to its value in an unbounded fluid. The abscissa is the distance from the boundary divided by the bubble radius. **b** Variation of the phase for the bubble rising near the rigid bottom (*solid line*) and a free bubble (*dashed line*), respectively. An asterisk indicates a point of time when the phase is a multiple of 2π. Time is measured in periods of natural oscillations of the free bubble

the bubble will rise during radiation time (about 25 cycles) at distances comparable with its radius. Thus the radiation comes from a region immediately adjacent to the bottom.

The variation (decreasing) of the period of the bubble's natural oscillations with the increasing distance to the wall is most clearly manifested in the changing intervals between consecutive zeros of the pressure recorded by the hydrophone. In order to assess these changes, we calculated the phase dependence of bubble oscillations on time and compared it with the phase variation of a free bubble

$$\Phi(t) = \int_0^t dt' \Omega_*(t')$$
$$= 2\pi \int_0^\tau d\tau' \left\{ 1 - 2\sqrt{\kappa^2(\tau') - 1} \sum_{n=0}^{\infty} \frac{\left[\kappa(\tau') + \sqrt{\kappa^2(\tau') - 1}\right]^{-(2n+1)}}{\left[\kappa(\tau') + \sqrt{\kappa^2(\tau') - 1}\right]^{(2n+1)} + 1} \right\}^{1/2} \quad (7)$$

where time is measured in periods of natural oscillations of the free bubble: $\tau = t/T_0 = t\Omega_0/(2\pi)$. Figure 1b illustrates this dependence for the bubble of radius $R_0 = 0.5$ cm, rising at the speed $u = 25$ cm/s.

During the emission time (about 25 cycles), the bubble rises up at a relatively short distance ($\sim 3R_0$), so that a change in the period of its oscillation is small. The points in time when the phase becomes a multiple of 2π are marked with asterisks in Fig. 1b. It seems that this difference is difficult to identify in the field experiments due to the low level of the radiated signals and the relatively high level of the ambient noise in the ocean [7]. On the other hand, the technique used in the precise laboratory experiments [8, 9] has an adequate accuracy to detect these variations. One should note that even in the laboratory experiments where medical-grade breathing air was directed through copper tubing into a 10-cm long, 26 gauge, vertically oriented, stainless steel needle placed at the bottom of the tank, aspects of the acoustic signal radiated by a new bubble are dependent on specifics of the orifice. For example, the amplitude of the radiated acoustic signal depends on the orientation of the orifice [9].

This theoretical study was initiated by a series of field experiments in the coastal area of the Sea of Japan and we present here the results of preliminary analysis.

4 Comparison with the Field Experiment

Observations of seabed bubbles at artificial marine seeps by passive acoustic and optical approaches have been carried out at the Vityaz Bay, Sea of Japan. Bubbles were generated at depths of 2, 6, and 10 m with an adjustable supply of compressed air through the various nozzles. The most interesting results were obtained when the air supply was made through the sandy layer thickness of a few centimeters.

Flow regimes in saturated granular media can go from homogeneous seepage (percolation) to piping and partial failure [10]. At homogeneous seepage, the grains form a skeleton through which the gas percolates homogeneously in upward direction. The porosity remains unchanged within the whole layer. Above a threshold pressure, which depends mainly on the grain size and on the surface free energy of the liquid–gas interface, the gas creates a channel within the whole thickness of the layer. The channel almost extends over the layer, and is sealed by a typical blockage at the top [11]. After emission of a bubble, only a small amount of granular material (typical size 10 bead diameters) at the upper end of the channel collapses and forms a blockage that seals the system.

As has been shown by Varas et al. [12], the spatial distribution of the bubble emissions at the free surface of an immersed granular bed can be accounted for by a diffusion like process. The typical size of the region in which the bubbles are emitted is directly related to the depth of the granular bed h_g and scales like $\sqrt{Dh_g}$ [13], where the coefficient D scales like the typical grain size. A two-dimensional model, based on experimental observations [13], showed that the typical height and width of the region explored by the branched paths depend not only on the injected volume V but also on a dimensionless parameter which accounts for the relative effects of the gravity and capillarity. In the discussed field experiments the camera recorded only the front view of the bubble. The particulate composition of sand was not studied. For this reason, we cannot compare the distribution of the bubble emissions at the surface of the sand bed with the proposed law [13].

A camera with a shutter speed of 1/1000 s was placed horizontally in front of a container with sediments to record uncompressed video at a rate of 30 frame/s. The optical system provided insight into the shapes and behaviors of the bubbles. The ruler was used as a size reference. However, the method of direct sizing using images carried with it several potential sources of error. The bubble appears blurry, likely due to the bubble pulsations and rising. Optical measurements were plagued by uncertainties resulting from brightness thresholding as it affected the determination of bubble outlines.

The hydrophone signal was conditioned with an amplifier then band-pass filtered before being captured by the data acquisition system. The typical FFT of recorded signal of birthing bubble is shown in Fig. 2. Unexpected feature, which was not observed in laboratory experiments with bubbles created at the tip of the steel needle, is the presence of a marked component in the subharmonic range. Analysis of the acoustic signals recorded in the field experiments in the Sea of Japan and attempts to compare them with the predictions of the theory did not lead to an unambiguous result for bubbling through the sandy layer. However, a preliminary analysis shows that the model of the rigid bottom is not fully suitable for describing the acoustic emission of bubbles birthing in the sandy sediments.

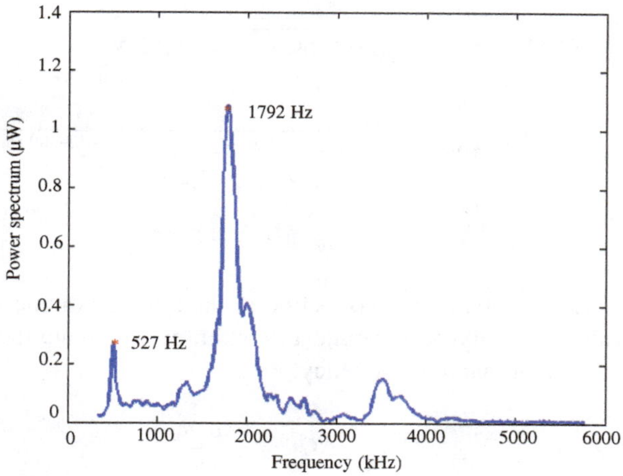

Fig. 2 The power spectrum corresponding to the signal picked up by the hydrophone at the distance of near 30 cm from the modest artificial seep at the depth of 6 m

5 Impedance Boundary Condition

The simplest generalization is the impedance boundary condition which can account for the elasticity of the seabed:

$$-\mathrm{i}\omega\rho_1\varphi = w\frac{1}{h_\xi}\left(\frac{\partial\varphi}{\partial\xi}\right)_{\xi=0} \tag{8}$$

This condition properly accounts for the limiting cases of the rigid $w \to \infty$ and the soft $w \to 0$ boundaries. We will consider the obtained solution for the rigid boundary (5) as the main approximation $\varphi^{(0)}$ to the impedance solution, and will seek the correction $\varphi^{(1)}$, assuming smallness of the parameter $\varepsilon \equiv \omega\rho_1 a/|w| \ll 1$. Thus the boundary condition (8) takes the form as follows:

$$-\mathrm{i}\omega\rho_1\varphi^{(0)} = w\frac{1}{h_\xi}\left(\frac{\partial\varphi^{(1)}}{\partial\xi}\right)_{\xi=0} \tag{9}$$

The general solution of the Laplace equation, vanishing on the surface of the bubble ($\xi = \xi_0$) has the form

$$\varphi^{(1)}(\xi,\vartheta,t) = \varphi_\infty(t)\sqrt{\cosh\xi - \cos\vartheta}\sum_{n=0}^{\infty} A_n \frac{\sinh[(n+1/2)(\xi-\xi_0)]}{\cosh[(n+1/2)\xi_0]} P_n(\cos\vartheta) \tag{10}$$

where the coefficients A_n are determined by the boundary condition Eq. (9). Direct, but rather cumbersome calculations lead to the following result:

$$\begin{aligned}\varphi^{(1)}(\xi,\vartheta,t) &= \varphi_\infty(t)\sqrt{\cosh\xi-\cos\vartheta}\sqrt{2}\sum_{n=0}^{\infty}\frac{e^{-(n+1/2)\xi_0}\cosh[(n+1/2)\xi]}{\sinh[(n+1/2)\xi_0]}\\ &\times y_n^{(1)}\left[\frac{1}{1+e^{-(2n+1)\xi_0}}-\frac{1}{1+e^{-(2n+1)\xi_0}}\right]P_n(\cos\vartheta) \qquad (11)\\ y_n^{(1)} &= -\frac{i\omega\rho_1 a}{w}\sum_{n'=0}^{n-1}\frac{1}{(n'+1)}\sum_{n''=0}^{n'}(n''+1)\tanh[(n''+1/2)\xi_0]\end{aligned}$$

By calculating the correction to the bubble volume $\Delta V^{(1)}$, and substituting this value in the dynamic boundary condition, we obtain the correction to the square of the bubble natural frequency:

$$\Omega_*^2=\Omega_0^2(\tilde{\Omega}^2+\Delta\tilde{\Omega}^2),\quad \Delta\tilde{\Omega}^2=\frac{i\omega\rho_1 R_0}{w}F,$$

$$F=8\sqrt{2}\sum_{n=1}^{\infty}\frac{\left[\kappa+\sqrt{\kappa^2-1}\right]^{-(2n+1)}}{\left[\kappa+\sqrt{\kappa^2-1}\right]^{(2n+1)}+1}\sum_{n'=0}^{n-1}\frac{1}{n'+1}\sum_{n''=0}^{n'}\frac{\left[\kappa+\sqrt{\kappa^2-1}\right]^{-(2n''+1)}}{\left[\kappa+\sqrt{\kappa^2-1}\right]^{(2n''+1)}+1} \qquad (12)$$

The real R and imaginary X parts of $w = R + iX$ are the specific resistance and reactance, respectively. In a wave that is incident upon a reflecting interface surface, such as a material discontinuity, the magnitude of the ratio of the amplitude of the reflected wave to the amplitude of the incident wave is less than 1 if the real part of w is positive. Any surface having this property absorbs acoustic energy. Thus, the resistive part is responsible for the bubble radiation damping. The reactive part represents the ability of sediment to store the kinetic energy of the wave as potential energy. For reactive effects the energy is not lost from the system, but converted between kinetic and potential forms.

When a plane wave is incident upon a flat liquid/solid interface at the incidence angle θ_i, the normalized impedance $w_n = w/(\rho_1 c_1)$ is described by the following formula:

$$\begin{aligned} w_n &= \frac{\rho_2 c_{l2}}{\rho_1 c_1}\frac{\cos^2 2\theta_l}{\cos\theta_l}+\frac{\rho_2 c_{t2}}{\rho_1 c_1}\frac{\cos^2 2\theta_t}{\cos\theta_t} \qquad (13)\\ \cos\theta_l &= \sqrt{1-(c_{l2}/c_1)^2\sin^2\theta_i},\quad \cos\theta_t=\sqrt{1-(c_{t2}/c_1)^2\sin^2\theta_i}\end{aligned}$$

where the subscripts l and t refer to longitudinal and transverse waves in marine sediments. For more complex perturbations, in particular, for the velocity components of an incompressible flow field, the angle θ_i should be defined so as to be valid not only for the real values, but also for complex ones [14]. Relationship between the phenomenological parameters R, X and an exact integral representation of the problem of diffraction (by an elastic sphere near an elastic half-space) in the form of a contour integral along a path on the complex angles plane [14] is in an approximation of the variable impedance Eq. (13) by its characteristic value at $|\cos\theta_i| \sim c_1/(\Omega_0 h)$.

An important fact is that the correction to the natural frequency due to reactance is positive thereby reducing the influence of the rigid boundary

$$\Delta\tilde{\Omega}^2_{\text{react}} = \frac{\Omega_0 \rho_1 R_0 X}{R^2 + X^2} F \tag{14}$$

Thus we have shown that the presence of soft (gas filled) channels in the sand can be taken into account within a simple model of the impedance boundary, and thus explain the absence of a noticeable shift in the natural frequency observed in the experiment.

6 Discussion and Conclusions

When gas leaks occur underwater, bubbles are produced and emit sound at frequencies intimately related to their sizes. The important thing is that, during the emitting time, the bubble is located near the point where it was produced. As a result, the inertial mass of bubble oscillations depends on the material of the boundary. Thus the note produced by a bubble depends not only on its size, but on the environment: the steel pipe or elastic sediments. Moreover, the rising gas creates channels within the whole thickness of the upper sediment layer. The presence of these gas-filled channels has been accounted for in the current study in the frame of a simple phenomenological model of the impedance boundary condition. The next step is a more sophisticated model accounting for coupling oscillations of a birthing bubble and a gas pocket on a liquid-covered solid surface. The theory of natural oscillations of such a pocket has been derived in the recent study [15].

Subsea leak detection is becoming increasingly important in the gas and petroleum industry. The problem for sensor performance is unwanted warnings, often generated due to natural seepage of hydrocarbons from the seabed. The existence of the principal ability to discriminate sounds produced by leakage of a stile pipe and natural seeps allows one to hope for the solution of this problem.

Acknowledgments The study was supported by RFBR 11-05-00212a, MES SS 1052.2012.5, and FEBRAS 12-III-A-07-131, 12-III-A-07-124.

References

1. Leighton, T.G., White, P.R.: Quantification of undersea gas leaks from carbon capture and storage facilities, from pipelines and from methane seeps, by their acoustic emissions. In: Proceedings: Mathematical, Physical and Engineering Sciences, vol. 468, pp. 485–510 (2012)
2. Leighton, T.G.: The Acoustic Bubble. Academic Press, London (1994)

3. Maksimov, A.O., Burov, B.A., Salomatin, A.S., Chernykh, D.V.: On sounds of marine seeps. In: Proceedings of the 1st Underwater Acoustics International Conference & Exhibition, pp. 827–834 (2013)
4. Oguz, H.N., Prosperetti, A.: Bubble oscillations in the vicinity of a nearly plane free surface. J. Acoust. Soc. Am. **87**(5), 2085–2092 (1990)
5. Maksimov, A.O.: On the volume oscillations of a tethered bubble. J. Sound Vibr. **283**, 915–926 (2005)
6. Maksimov, A.O., Polovinka, Y.A.: Volume oscillations of a constrained bubble. Phys. Fluids **25**, 062104 (2013)
7. Leifer, I., Tang, D.: Characteristics of acoustic scattering from hydrate shelled bubbles. J. Acoust. Soc. Am. **121**(1), 35–40 (2007)
8. Payne, E.M.B., Illesinghe, S.J., Ooi, A., Manasseh, R.: Symmetric mode resonance of bubbles attached to a rigid boundary. J. Acoust. Soc. Am. **118**(5), 2841–2849 (2005)
9. Green, C.A., Wilson, P.S.: Laboratory investigation of a passive acoustic method for measurement of underwater gas seep ebullition. J. Acoust. Soc. Am. **131**(1), 61–66 (2012)
10. Wilhelm, T., Wilmanski, K.: On the onset of flow instabilities in granular media due to porosity inhomogeneities. Int. J. Multiphase Flow **28**, 1929–1944 (2002)
11. Gostiaux, L., Gayvallet, H., Géminard, J.C.: Dynamics of a gas bubble rising through a thin immersed layer of granular material: an experimental study. Granular Matter **4**, 39–44 (2002)
12. Varas, G., Vidal, V., Géminard, J.C.: Venting dynamics of an immersed granular layer. Phys. Rev. **E 83**, 011302 (2011)
13. Varas, G., Vidal, V., Géminard, J.C.: Morphology of air invasion in an immersed granular layer. Phys. Rev. **E 83**, 061302 (2011)
14. Shenderov, E.L.: Diffraction of sound by an elastic or impedance sphere located near an impedance or elastic boundary of a halfspace. Phys. Acoust. **48**, 607–617 (2002)
15. Gelderblom, H., Zijlstra, A.G., van Wijngaarden, L., Prosperetti, A.: Oscillations of a gas pocket on a liquid-covered solid surface. Phys. Fluids **24**, 122101 (2013)

Multi-AUV Localization for an Underwater Acoustic Sensor Network

Bo Li, Yuanxin Xu and Wen Xu

Abstract In this paper, a new approach for localizing underwater sensor nodes is presented. Multi-autonomous underwater vehicles (AUVs) are used in our method (in this paper we use four AUVs). Similar to a global satellite navigation system, each AUV broadcasts its position and the time when it sends its messages. Nodes in the communication range can receive the messages and detect the range differences from the four AUVs to the sensor nodes. Then we calculate their positions through trilateration by the range differences and the AUVs' positions. Finally, both theoretical analysis and simulation results are given to confirm our method. There are many advantages in our approach: (1) it requires no time synchronization between nodes, but only requires time synchronization between the AUVs which can be easily achieved; (2) each AUV can run along a line, which makes the network nodes localization very fast and effective, and it is also easy for path planning; (3) one node can receive more than one message from one AUV, and we can get the node's position more accurately.

Keywords Underwater acoustic sensor network (UASN) · Range-based scheme · Root mean square error (RMSE)

1 Introduction

An underwater acoustic sensor network (UASN) can perform collaborative monitoring tasks over a given area with a variable number of sensors. It has emerged as a new alternative technology that can offer significant advantages and benefits in

B. Li (✉) · Y. Xu · W. Xu
College of Information Science and Electronic Engineering,
Zhejiang University, Hangzhou 310027, China
e-mail: libo3071102052@zju.edu.cn

Y. Xu
e-mail: xuyx@zju.edu.cn

W. Xu
e-mail: wxu@zju.edu.cn

L. Zhou et al. (eds.), *Underwater Acoustics and Ocean Dynamics*
DOI 10.1007/978-981-10-2422-1_16

many applications, such as navigation assistance, target detection, underwater environmental observation, tactical surveillance, commercial exploitation, and disaster prevention, etc.

Sensor network data is commonly used with a sensor's location, so location discovery for underwater sensor nodes is very important in the oceanic medium and it is also a challenging task for UASNs. Because the magnetic fading is fast in the water, the Global Positioning System (GPS) receivers cannot work properly in underwater [1]. In a UASN we need to get the node's position in a 3-D space. In order to save energy, we should have minimal message exchange. Different form the terrestrial networks, underwater networks use acoustic communications. In essence, the major challenges of localization in UASNs arise from physical layer constraints. Acoustic channels have multipath interference and high propagation delays. The speed of sound varies with pressure, temperature, and salinity which can affect the localization accuracy [2].

Lots of localization schemes have been studied, and they commonly can be classified into two categories: range-based schemes, and range-free schemes [3]. In the range-based schemes, angles or distance measurements are needed for estimating the location of nodes in underwater networks. The time of arrival (TOA), angle of arrival (AOA) or received signal strength indicator (RSSI), and time difference of arrival (TDOA) are used to estimate their distances to other nodes. Range-free localization schemes do not use any range or bearing information. The DV-Hop [4] and Density aware Hop-count Localization (DHL) [5] are range-free localization, and the advantages of these schemes lie in their simplicity. Range-based schemes always have higher position estimation accuracy than range-free schemes.

Range-based schemes are potentially a good choice for UASNs. There are some studies on localization for UASNs in Refs. [6, 7, 8]. In Ref. [6], a silent positioning scheme is introduced by Cheng and it is effective and does not need time synchronization. However, it is just for one-hop underwater networks and it is based on four static buoys, which limits the localization areas. In Ref. [7], the 3-D underwater positioning problem is transformed into a 2-D counterpart using the projection technique, and it can greatly simplify the localization process. In Ref. [8], the authors use mobile beacon nodes to aid the localization process, and the AUV works as a mobile satellite to localize the nodes within its communication range. However, this scheme needs the nodes' responses which may utilize a great deal of energy and it may also require time synchronization between the AUV and all the nodes. In our approach, there is no requirement for time synchronization between the nodes, the only requirement is for time synchronization between the AUVs which can be easily achieved. Each node only receives signals, and does not need to send any response messages to the AUVs which can save a great deal of energy for the nodes. Each AUV can run along a straight line, and it makes the sensor nodes localization very fast and effective. The movement of the AUVs also makes the localization area flexible and large.

The remainder of this paper is organized as follows: Sect. 2 proposes our localization approach. Theoretical performance analysis is given in Sect. 3. Section 4 provides the simulation results. We conclude this paper in Sect. 5 with a discussion of future research directions.

2 Multi-AUV Localization Scheme

In this section, our localization approach is introduced: a Multi-AUV localization scheme (MLS). First the network architecture and an overview of MLS are described. Then its details are shown.

MLS consists of two phases. Phase 1: AUVs get time synchronization and nodes receive the messages from the AUVs. The differences of the signal arrival times between the nodes and the four AUVs are detected, and then the nodes calculate the range differences. Phase 2: Nodes estimate their positions through trilateration using the AUVs' positions and the range differences.

2.1 Network and Scheme Model

We assume that there are only stationary sensors in the UASN. The network architecture is ocean floor embedded with sensors or suspending sensors. There exist four AUVs whose locations are known exactly. Each of AUVs is equipped with an acoustic transmitter that has a certain communication range. Each AUV runs through the network and broadcasts messages every τ s, and the messages include the AUV's position and the time when it sends the messages. The nodes in the communication range will receive the messages and locate themselves. The layout of the scheme is shown in Fig. 1.

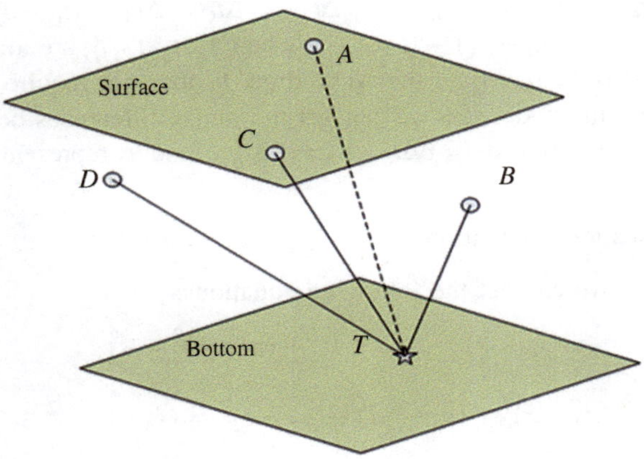

Fig. 1 Four AUVs named *A*, *B*, *C*, and *D*. *C* and *A* are at the surface; *B* and *D* are at the same depth underwater. Sensor *T* is at the bottom or suspending in water and will receive messages from the AUVs. The AUVs' messages will be repeated every τ s. The messages include each AUV's position and the time it sends messages

2.2 Range-Based Localization

We assume the four AUVs are A, B, C, and D. Given the locations of four AUVs, $(x_a, y_a, z_a), (x_b, y_b, z_b), (x_c, y_c, z_c), (x_d, y_d, z_d)$, respectively, and z represents the depth. Assume the target sensor is T and the target's location (x, y, z) is going to be determined, as in Fig. 1. Assume the distance between i and T is s_{iT}, where $i \in \{A, B, C, D\}$, as shown in Fig. 1, representing the four AUVs. We have the following equations:

$$(x_a - x)^2 + (y_a - y)^2 + (z_a - z)^2 = s_{AT}^2 \tag{1}$$

$$(x_b - x)^2 + (y_b - y)^2 + (z_b - z)^2 = s_{BT}^2 \tag{2}$$

$$(x_c - x)^2 + (y_c - y)^2 + (z_c - z)^2 = s_{CT}^2 \tag{3}$$

$$(x_d - x)^2 + (y_d - y)^2 + (z_d - z)^2 = s_{DT}^2 \tag{4}$$

- Phase 1: Time synchronization and range difference calculation

First, we let the four AUVs receive GPS signals to locate themselves and obtain time synchronization. Then the AUVs dive into certain depths and follow a predefined trajectory, moving among the sensor nodes. Finally, each AUV broadcasts messages every τ s, and assume that the AUVs broadcast at the time t_A, t_B, t_C, t_D. The target node receives the messages at t'_A, t'_B, t'_C, t'_D, and the time from A to T is $\Delta t_A = t'_A - t_A$. The time from B, C, D to T is $\Delta t_B = t'_B - t_B$, $\Delta t_C = t'_C - t_C$, $\Delta t_D = t'_D - t_D$, so the time difference between Δt_A and $\Delta t_B, \Delta t_C, \Delta t_D$ is $\Delta t_{AB} = \Delta t_B - \Delta t_A = (t'_B - t'_A) - (t_B - t_A)$, $\Delta t_{AC} = \Delta t_C - \Delta t_A = (t'_C - t'_A) - (t_C - t_A)$, $\Delta t_{AD} = \Delta t_D - \Delta t_A = (t'_D - t'_A) - (t_D - t_A)$, where t'_A, t'_B, t'_C, t'_D are local times at the target node, and t_A, t_B, t_C, t_D are the AUV time. It does not require time synchronization between the nodes. So we can get the range differences between AT and BT, CT, DT. $a_1 = c\Delta t_{AB}, a_2 = c\Delta t_{AC}, a_3 = c\Delta t_{AD}$, and c represents the acoustic speed underwater.

- Phase 2: Location calculation

From Phase 1 we can get the following equations:

$$(x_a - x)^2 + (y_a - y)^2 + (z_a - z)^2 = s_{AT}^2 \tag{5}$$

$$(x_b - x)^2 + (y_b - y)^2 + (z_b - z)^2 = (s_{AT} + a_1)^2 \tag{6}$$

$$(x_c - x)^2 + (y_c - y)^2 + (z_c - z)^2 = (s_{AT} + a_2)^2 \tag{7}$$

$$(x_d - x)^2 + (y_d - y)^2 + (z_d - z)^2 = (s_{AT} + a_3)^2 \tag{8}$$

Solving these equations, we obtain

$$x = A_x s_{AT} + B_x \tag{9}$$

$$y = A_y s_{AT} + B_y \tag{10}$$

$$z = A_z s_{AT} + B_z \tag{11}$$

$$s_{AT} = \frac{-b \pm \sqrt{b^2 - 4ac}}{2a} \tag{12}$$

where

$$a = A_x^2 + A_y^2 + A_z^2 - 1, \quad b = 2(A_x k_1 + A_y k_2 + A_z k_3), \quad c = k_1^2 + k_2^2 + k_3^2,$$
$$k_1 = B_x - x_a, \quad k_2 = B_y - y_a, \quad k_3 = B_z - z_a,$$
$$A_x = b_{11}\alpha_1 + b_{12}\alpha_2 + b_{13}\alpha_3, \quad A_y = b_{21}\alpha_1 + b_{22}\alpha_2 + b_{23}\alpha_3, \quad A_z = b_{31}\alpha_1 + b_{32}\alpha_2 + b_{33}\alpha_3,$$
$$B_x = b_{11}\beta_1 + b_{12}\beta_2 + b_{13}\beta_3, \quad B_y = b_{21}\beta_1 + b_{22}\beta_2 + b_{23}\beta_3, \quad B_z = b_{31}\beta_1 + b_{32}\beta_2 + b_{33}\beta_3,$$
$$\alpha_1 = 2a_1, \quad \alpha_2 = 2a_2, \quad \alpha_3 = 2a_3, \quad \beta_1 = a_1^2 + x_a^2 - x_b^2 + y_a^2 - y_b^2 + z_a^2 - z_b^2,$$
$$\beta_2 = a_2^2 + x_a^2 - x_c^2 + y_a^2 - y_c^2 + z_a^2 - z_c^2, \quad \beta_3 = a_3^2 + x_a^2 - x_d^2 + y_a^2 - y_d^2 + z_a^2 - z_d^2,$$
$$b_{ij} = A^{-1}(i,j), \quad A = 2\begin{bmatrix} x_a - x_b & y_a - y_b & z_a - z_b \\ x_a - x_c & y_a - y_c & z_a - z_c \\ x_a - x_d & y_a - y_d & z_a - z_d \end{bmatrix}$$

If $s_{AT} > 0$ is unique, then we can locate the target sensor, and we call the target area a feasible area. In Sect. 4, we will give some simulations for the feasible space. In our scheme, one node may receive more than one message from one AUV, and then the node's positions are calculated by averaging all of the estimated positions. Therefore, we can get more accurate location estimations.

3 Theoretical Performance Analysis

In this section, the position error of MLS that resulted from the AUVs' positions are studied. There are four sources of errors in our position scheme: the AUVs positions error, the time synchronization error, the variable acoustic speed underwater, and the underwater multipath fading. In this paper, we just consider the most important part: the AUVs' position errors.

In our approach, all errors are finally included in a_1, a_2, a_3, so we just consider the range differences error. Because the z position of the AUV comes from the depth sensor, its error does not have a cumulative effect, and the z position can have a high accuracy. We assume that the AUV's z position has no error, and the x, y position error is $\omega \sim N(0, \sigma^2)$. Through some derivations, we get the $\Delta a_1, \Delta a_2, \Delta a_3$ approximate distributions as follows:

$$\Delta a_1 \sim N(0, \sigma_{a1}^2), \quad \Delta a_2 \sim N(0, \sigma_{a2}^2), \quad \Delta a_3 \sim N(0, \sigma_{a3}^2)$$

where

$$\sigma_{a1}^2 = \sigma^2\left(2 - \frac{(z_b - z_0)^2}{s_{BT}^2} - \frac{(z_a - z_0)^2}{s_{AT}^2}\right), \quad \sigma_{a2}^2 = \sigma^2\left(2 - \frac{(z_c - z_0)^2}{s_{CT}^2} - \frac{(z_a - z_0)^2}{s_{AT}^2}\right)$$

$$\sigma_{a3}^2 = \sigma^2\left(2 - \frac{(z_d - z_0)^2}{s_{DT}^2} - \frac{(z_a - z_0)^2}{s_{AT}^2}\right)$$

where $s_{AT}, s_{BT}, s_{CT}, s_{DT}$ are the ranges between A, B, C, D and the target sensor T, z_0 is the target sensor's depth.

Then we can get the mean and variance of the estimated position as follows:

$$\mu_x \approx \mu_y \approx \mu_z \approx 0 \tag{13}$$

$$Var(x) \approx 4R_a^2(b_{11}^2\sigma_{a1}^2 + b_{12}^2\sigma_{a2}^2 + b_{13}^2\sigma_{a3}^2) \tag{14}$$

$$Var(y) \approx 4R_a^2(b_{21}^2\sigma_{a1}^2 + b_{22}^2\sigma_{a2}^2 + b_{23}^2\sigma_{a3}^2) \tag{15}$$

$$Var(z) \approx 4R_a^2(b_{31}^2\sigma_{a1}^2 + b_{32}^2\sigma_{a2}^2 + b_{33}^2\sigma_{a3}^2) \tag{16}$$

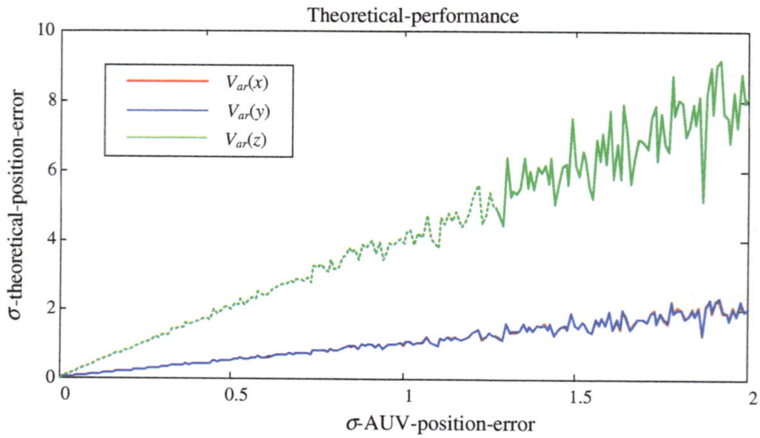

Fig. 2 Four AUVs' positions: $A(-100, 100, -50)$, $B(100, 100, 0)$, $C(100, -100, -50)$, $D(-100, -100, 0)$. The target's true position is: $T(40, -30, -200)$

The variables in these equations have the same meanings as those mentioned before. We can find that the estimated position errors are dependent on the target's location.

Figure 2 shows the relationship between the AUVs' position errors and the estimated sensor position error. We can find that the relationship is approximately linear.

The variable acoustic speed underwater and the underwater multipath fading may have a great effect on the estimation in shallow water. For more details see Ref. [6].

4 Simulation

In this section, the performance of our approach is studied.

First, we check out the feasible space when AUVs are at different structures. If all AUVs are in the same plane, Eqs. (5)–(8) will have no solutions and we cannot locate the target sensor. If $s_{AT} > 0$ is unique, we call the target sensor feasible. Figure 3 represents the feasible areas of different AUV structures.

We can see that the network localization coverage is higher if the target sensor is near the underwater AUVs, and if the AUVs go through *AC* or *CA*, we can get the

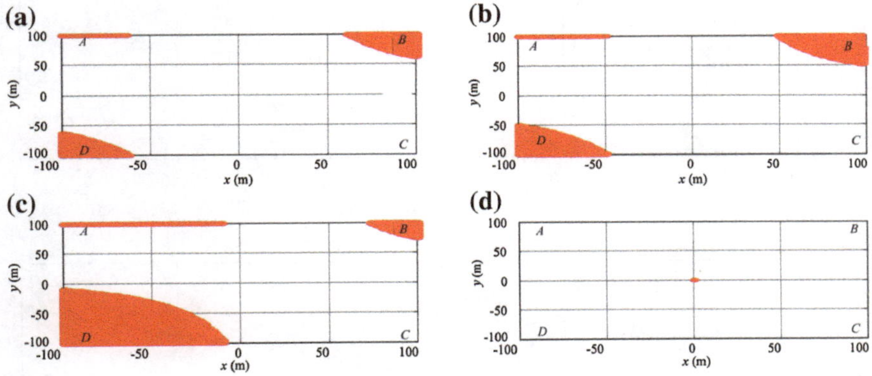

Fig. 3 **a** The network localization coverage is 95.47 %; the locations of the four AUVs are *A*(−100, 100, 0), *B*(100, 100, −100), *C*(100, −100, 0), *D*(−100, −100, −100). The target sensor *T* is at the depth of 200 m. **b** The network localization coverage is 91.86 %; the locations of the four AUVs are *A*(−100, 100, 0), *B*(100, 100, −50), *C*(100, −100, 0), *D*(−100, −100, −50); the target sensor *T* is at the depth of 200 m. **c** The network localization coverage is 84.02 %; the locations of the four AUVs are *A*(−100, 100, 0), *B*(100, 100, 0), *C*(100, −100, 0), *D*(−100, −100, −50); the target sensor *T* is at the depth of 200 m. **d** The network localization coverage is 100 %; the locations of the four AUVs are *A*(−100, 100, 0), *B*(100, 100, −200), *C*(100, −100, 0), *D*(−100, −100, −200); the target sensor *T* is at the depth of 100 m

network localization coverage of 100 %. In our approach, the AUVs can go along a straight line, which makes the localization fast and effective. Also if an AUV runs along a straight line, its position accuracy is the highest for current AUV system model.

Then we consider the AUVs' position error. Assume that the x, y position has an error $\omega \sim N(0, \sigma^2)$ and z position has no error.

From Fig. 4, we can find that there is an approximately linear relationship between the AUVs' position errors and the estimated target sensor errors which is the same as the theoretical analysis. In Fig. 4a the estimated position error of z is much larger than the errors of x, y in this case, because the AUV's depth is too far away from the target sensor. Figure 4b represents the case that the target resides in the enclosed space for the four AUVs. In order to get a more accurate estimation, we can make the target near the AUVs or the target resides in the enclosed space for the four AUVs, and also we need highly accurate AUVs' positions. In our schemes, four AUVs get out of the water to achieve their locations assignment and time synchronization occurs at every τ min.

If the target node receives more than one message from one AUV, we can get the estimated positions by averaging all the results and it can decrease the variance of the estimated errors. Figure 5a represents the case that the target sensor is the outside of the enclosed space for the four AUVs, and Fig. 5b represents the case

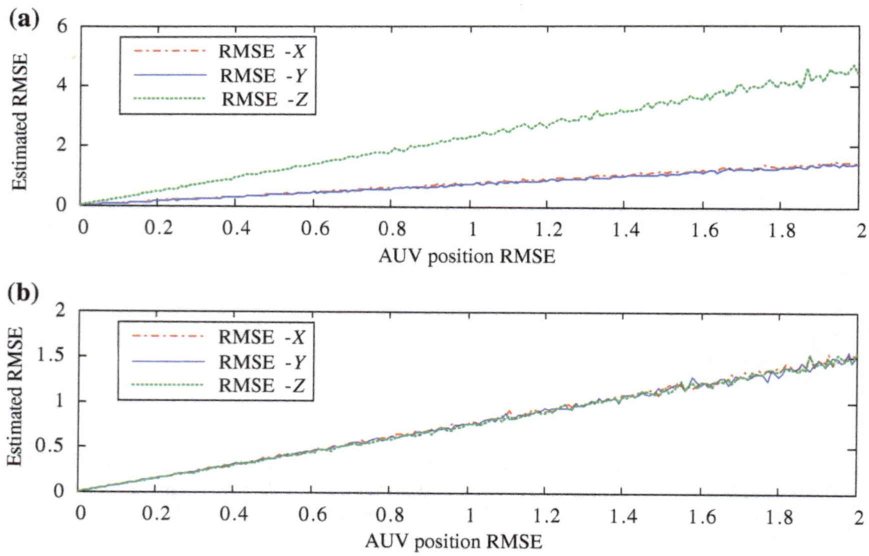

Fig. 4 **a** Locations of the four AUVs are $A(-100, 100, -50)$, $B(100, 100, 0)$, $C(100, -100, -50)$, $D(-100, -100, 0)$; the target sensor is located at $T(40, -30, -200)$. **b** Locations of the four AUVs are $A(-100, 100, -200)$, $B(100, 100, 0)$, $C(100, -100, -200)$, $D(-100, -100, 0)$; the target sensor is located at $T(40, -30, -100)$

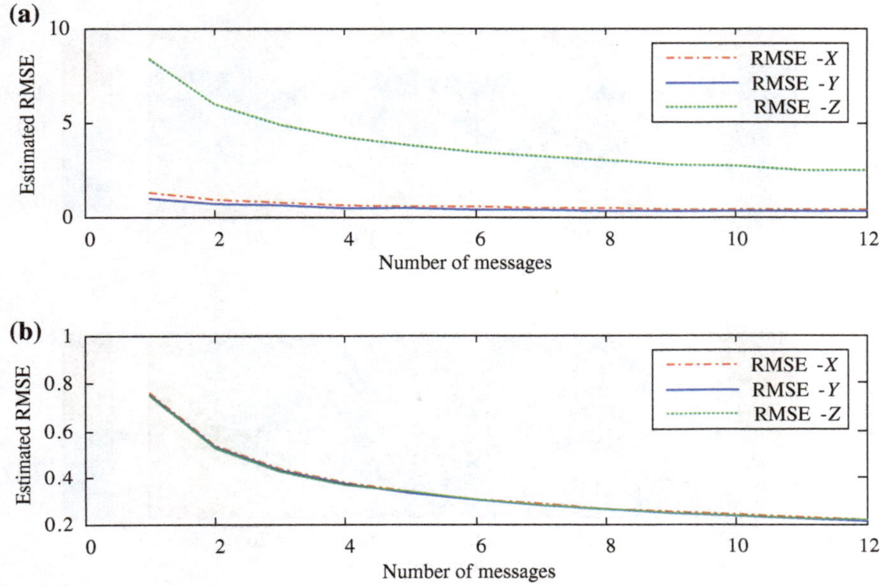

Fig. 5 **a** Locations of the four AUVs are $A(-100, 100, -50)$, $B(100, 100, 0)$, $C(100, -100, -50)$, $D(-100, -100, 0)$; The target sensor is located at $T(40, -30, -200)$. **b** Locations of the four AUVs are $A(-100, 100, -200)$, $B(100, 100, 0)$, $C(100, -100, -200)$, $D(-100, -100, 0)$; the target sensor is located at $T(40, -30, -100)$. The AUV position error is $\sigma = 1$

that the target resides in the enclosed space. From Fig. 5, we can find that the estimated position error decreases as the received messages increase.

From Eqs. (14)–(16), we can find that the estimated position errors are dependent on the target's true position. So we calculate the position errors versus the true target position, as shown in Fig. 7 and the four AUVs topology is shown in Fig. 6.

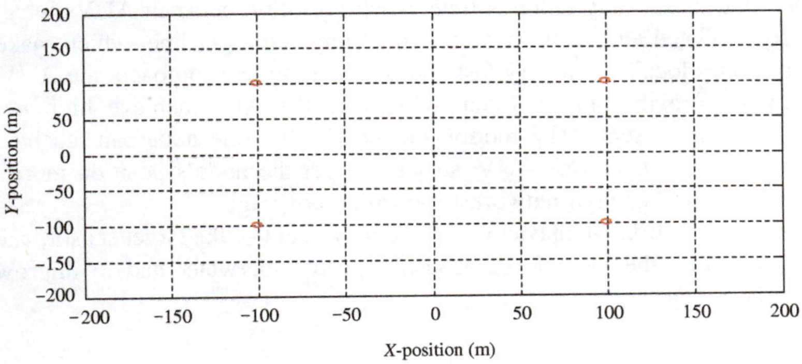

Fig. 6 Four AUVs topology structure

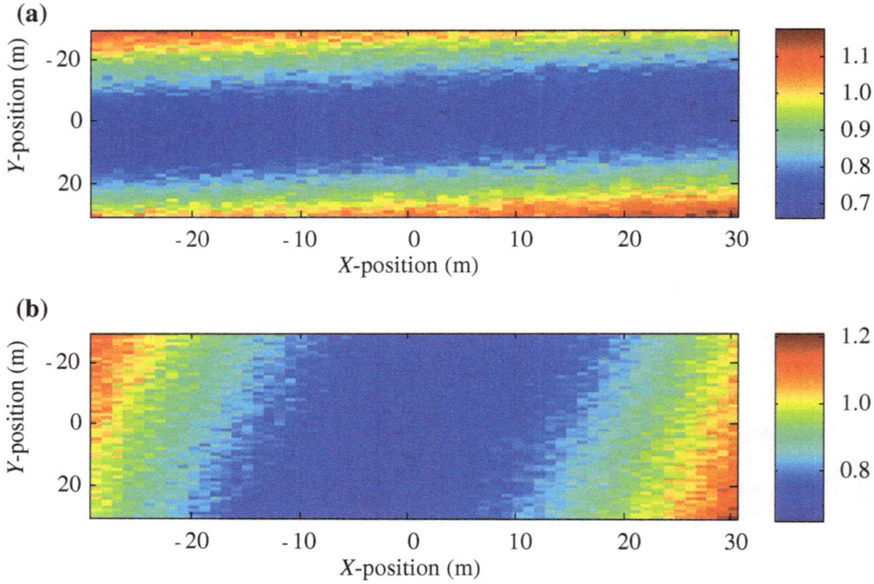

Fig. 7 Locations of the four AUVs are $A(-100, 100, -50)$, $B(100, 100, 0)$, $C(100, -100, -50)$, $D(-100, -100, 0)$; the target sensor is at the depth of 200 m. **a** The estimated *X*-position error versus the real target position; **b** the estimated *Y*-position error versus the real target position

From Fig. 7, we can find that if the target is along the *X*-axis, the *X*-position error is small, and also if the target is along the *Y*-axis, the *Y*-position error is small.

5 Conclusion and Future Work

There are many advantages in our approach. It requires no time synchronization between nodes, but only requires time synchronization between AUVs which can be easily achieved and each AUV can run along a straight line, which makes the network nodes localization very fast and effective. In our approach, the AUVs can run along a straight line, and that will make the AUV achieve high position accuracy in the current AUV motion model. Finally, one node can receive more than one message from one AUV, so we can get the node's position more accurately and it also has high network localization coverage.

In our future work, we may consider the movement of the target sensor, and also we will consider the variable of acoustic speed underwater and to improve the position accuracy.

References

1. Akyildiz, I.F., Pompili, D., Melodia, T.: Underwater acoustic sensor networks: Research challenges. J. Ad Hoc Netw. **3**(3), 257–279 (2005)
2. Heidemann, J., Ye, W., Wills, J., Syed, A., Li, Y.: Research challenges and applications for underwater sensor networking. In: Wireless Communications and Networking Conference, vol. 1, pp. 228–235 (2006)
3. Chandrasekhar, V., Seah, W.K.G., Choo, Y.S., Ee, H.V.: Localization in underwater sensor networks—survey and challenges. In: Proceedings of the 1st ACM International Workshop on Underwater Networks, pp. 33–40 (2006)
4. Niculescu, D., Nath, B.: DV based positioning in ad hoc networks. Telecommun. Syst. **22**(1–4), 267–280 (2003)
5. Wong, S.Y., Lim, J.G., Rao, S.V., Seah, W.K.G.: Multihop localization with density and path length awareness in non-uniform wireless sensor networks. In: Proceedings of the 61st IEEE Vehicular Technology Conference (VTC2005-Spring), Stockholm, Sweden (2005)
6. Cheng, X., Shu, H., Liang, Q., Du, D.H.C.: Silent positioning in underwater acoustic sensor networks. IEEE Trans. Veh. Technol. **57**(3), 1756–1766 (2008)
7. Erol, M., Vierira, L.F.M., Gerla, M.: AUV-aided localization for underwater sensor networks. In: Proceedings of International Conference on Wireless Algorithms, Systems and Applications (WASA), pp. 44–51 (2007)
8. Cheng, W., Teymorian, A.Y., Ma, L., Cheng, X., Lu, X., Lu, Z.: Underwater localization in sparse 3D acoustic sensor networks. In: INFOCOM 2008. The 27th Conference on Computer Communications. IEEE, pp. 236–240 (2008)

FSC
www.fsc.org
MIX
Papier | Fördert gute Waldnutzung
FSC® C083411

Kolibri 360 GmbH
Ferdinand-Jühlke-Straße 7
99095 Erfurt, Deutschland
produktsicherheit@kolibri360.de